SpringerBriefs in Biochemistry and Molecular Biology

More information about this series at http://www.springer.com/series/10196

Jessica Nicastro · Shirley Wong
Zahra Khazaei · Peggy Lam
Jonathan Blay · Roderick A. Slavcev

Bacteriophage Applications—Historical Perspective and Future Potential

Jessica Nicastro
School of Pharmacy
University of Waterloo
Waterloo, ON
Canada

Shirley Wong
School of Pharmacy
University of Waterloo
Waterloo, ON
Canada

Zahra Khazaei
Western University
London, ON
Canada

Peggy Lam
University of Waterloo
Waterloo, ON
Canada

Jonathan Blay
School of Pharmacy
University of Waterloo
Waterloo, ON
Canada

Roderick A. Slavcev
School of Pharmacy
University of Waterloo
Waterloo, ON
Canada

ISSN 2211-9353   ISSN 2211-9361   (electronic)
SpringerBriefs in Biochemistry and Molecular Biology
ISBN 978-3-319-45789-5   ISBN 978-3-319-45791-8   (eBook)
DOI 10.1007/978-3-319-45791-8

Library of Congress Control Number: 2016949574

Printed on acid-free paper

This Springer imprint is published by Springer Nature
The registered company is Springer International Publishing AG
The registered company address is: Gewerbestrasse 11, 6330 Cham, Switzerland

Contents

Chapter 1
Overview of Bacteriophage Lifecycles and Applications

1 Introduction

Bacteriophages (phages) are well-established bacteria-specific viruses whose discovery is credited to the independent and nearly simultaneous works of Twort (1915) and d'Hérelle (1917) (Summers 1999) in the early 20th century. Each of the researchers characterized phages as the pathogens of bacteria following the hint of much phage-like phenomena from the 19th and 20th centuries. The late 1930s and early 1940s represented the most significant era for phage research and its impact on biological research (Abedon and Thomas-Abedon 2010), including the research by the "Phage Group". This group included the work of Max Delbrück and other highly notable geneticists, including James Watson and Francis Crick (Abedon 2012a). The group quickly established that phage could be used for the treatment of bacterial infections, since called "phage therapy", and were so named. "Bacteriophage" translates to "bacteria eaters".

While phage biology and the study of phage genetics were of interest, it was phage therapy that and its antibacterial potential that was the primary driver for phage research (Hanlon 2007; Summers 2001). Phage therapy however, failed to match the anticipation of its initially envisioned potential, particularly in a time when the phage themselves were poorly characterized, and the approach was thwarted in favour of small molecules in the Western World in the 1950s (Kropinski 2006; Summers 2001).

Although phage-based therapeutics did not meet the expectations of their initial interest, they have played a crucial role in the study of genetics and molecular biology (Henry and Debarbieux 2012), including contributions to the understanding of organisms much more complex than the phage themselves (Campbell 2003; Goodridge 2010). As such, the study of phage may have actually set the stage for its own demise in medicine.

© The Author(s) 2016
J. Nicastro et al., *Bacteriophage Applications—Historical Perspective and Future Potential*, SpringerBriefs in Biochemistry and Molecular Biology, DOI 10.1007/978-3-319-45791-8_1

There has been a strong resurfacing of interest in phage beyond the field of phage therapy and particularly, in phage-based technologies (Citorik et al. 2014; Henry and Debarbieux 2012). Phages offer great potential and impact in the food and agriculture, biotechnology, global nutrient cycling, and human health and disease industries. Furthermore, advances in genetics, bacteriology and synthetic biology have opened many opportunities to further phage-based therapeutics. This chapter provides an overview of phage biology and genetics as the governing principles necessary for the consideration of phage toward phage-based medical applications discussed in this book.

2   Phage Infection and Life-Cycle

Phage infection begins when the virion attaches to its host cell (adsorption) as part of the multi-step process of infection. This is shortly followed by the translocation of phage DNA into the host cell and subsequent expression of the phage genome within the host (Abedon 2012b; Samson et al. 2013). The aftermath of phage infection will depend on the phage, the host, and the circumstances of infection.

Successful adsorption will result in one of four circumstances: (i) the phage lives and replicates to form progeny and the host dies via lytic infection; (ii) both the phage and the host bacterium live and propagate, as seen with lysogeny, imparted by the lysogenic cycle of temperate bacteriophages; (iii) The phage dies and/or does not produce progeny and the host lives, a typical result of infection of hosts encoding restriction endonuclease (Labrie et al. 2010) and/or CRISPR/cas systems (Jiang and Doudna 2015); and (iv) both the phage and the host die as a result of abortive infection system(s) (Olszowska-Zaremba et al. 2012).

2.1   Lytic Phage

Most phage undergo lytic phage infection cycles whereby daughter progeny are produced and released at the expense of host cell lysis and death. These are considered "productive infections" where infections quickly lead to the release of viral progeny. Once the viral genome enters the host cell, phage-encoded genes are expressed in the bacterial cytoplasm, the functions of which, take over host bacterial metabolism (Young 2014). Infecting phages then enter a latent period during which phage particles are assembled and, once a threshold number of virions are produced, phage gene products "holin" and "lysin" (for classical double-stranded DNA phages) are responsible for the destruction of the host cell wall and subsequent release of the phage progeny to the extracellular matrix and neighboring cells (Abedon 2012a; Olszowska-Zaremba et al. 2012; Young 2014).

2.2   Temperate Phage

Bacteriophages that possess the ability to be stably harbored within their host as a prophage, thereby lysogenizing the host, are referred to as temperate. Temperate phages have the ability to switch between the lytic and lysogenic cycles, often existing as a prophage integrated into the host chromosome, but possessing the capacity to enter the lytic cycle in response to host and/or other external danger signals (typically the host SOS response) (Mardanov and Ravin 2007; Roberts and Devoret 1983).

Lysogenic cycles are characterized by two features: (i) the prophage is replicated sufficiently to permit daughter host cells to inherit at least one copy of the phage's DNA; and (ii) infections are not productive in that no structural virions are produced, but rather replication occurs vertically *in tandem* with host replication and division. While integration is perhaps a more common route of lysogeny, a prophage can manifest extra-chromosomally as a stable low copy plasmid that is not integrated into the host genome. Integration requires an integrase, which binds homologous segments of phage and bacterial DNA, resulting in site-specific recombination (Abedon 2012a).

The switch from a lysogenic cycle to productive infection, or lytic cycle, is known as induction or derepression. Prior to induction the phage will only produce proteins needed to maintain lysogeny, normally a repressor(s), necessary to prevent expression of all genes involved in the virus's vegetative growth, but also capable to trigger induction upon receiving the appropriate host/environmental signal(s) (Mardanov and Ravin 2007; Roberts and Devoret 1983).

3   Phage Infection Stages

3.1   Phage Entry

Bacteriophage infection begins when the particle adsorbs to a specific surface or appendage site(s) on a bacterial host cell. This initial recognition process is highly specific and typically involves the specific interaction of a binding ligand as some structural component of the phage with a corresponding receptor(s) on the host cell. Phages of gram-negative bacteria may recognize the polysaccharide moieties (e.g. phage T4) and/or outer membrane proteins (e.g. phage λ, T4) (Gaidelyte et al. 2006; Morita et al. 2002; Randall-Hazelbauer and Schwartz 1973). Gram-positive phages normally attach to the cell surface polysaccharides of the host (Valyasevi et al. 1990) and once a phage adsorbs, phage DNA translocates into the host cytoplasm where phage gene expression and replication may then occur.

3.2 DNA Replication

Viral gene expression ensues once the phage genome has entered its host cell (Abedon and Thomas-Abedon 2010). The phage genes will typically encode the capacity to harvest the host and exploit its metabolism to express its own genes. The specifics of phage DNA replication and the expression of the DNA products depend on the phage infecting the cell and the conditions surrounding the infection, including the species and attributes of the host cell (Mcnerney et al. 2004). The phage genome will typically code for assembly products for the production of phage progeny and for the amplification of its own genome.

3.3 Phage Assembly

Bacteriophages have served as model systems of viral assembly for the last half-century. Similar to phage DNA replication, the formation of phage particles and their individual structures will differ with each phage strain. However, despite extensive genetic diversity between phage genomes, similarities remain in structure and viral life cycle between bacterial viruses. These viral particles are essentially made up of two components: nucleic acid and a protein shell or capsid. Formation of the particles requires specific protein-protein and protein-nucleic acid interactions in addition to a well-established set of conformational changes resulting from each of these interactions, all of which are specific to the phage strain type (Aksyuk and Rossmann 2011). Examples from each of the major classes of phage (by nucleic acid genome), dsDNA, ssDNA, and ssRNA phage are described below in terms of phage assembly:

3.3.1 Tailed (dsDNA) Phages

All tailed bacteriophages' host ranges are determined by the specialized tail organelle. Despite a number of distinct strategies, most phages contain cell binding receptor proteins that bind host cells and trigger DNA release from the head. Phages with tails are from the *Caudoviridae* family and can be characterized by tail morphology as either short tails (*Podoviridae*), long non-contractile tails (*Siphoviridae*) or long contractile tails (*Myoviridae*) (Abedon 2012a; Ackermann 2003). Tailed phages follow several distinct steps for phage assembly: (i) assembly of a prohead, or the shell of capsid protein with a portal allowing for (ii) the packaging of DNA using ATP energy, (iii) maturation of proheads and (iv) attachment of the neck and tail proteins or a preassembled tail (Ackermann 2007).

3.3.2 ssDNA Phages

Filamentous phages from the *Inoviridae* family, including M13, fd and f1, are male (F plasmid)-specific ssDNA phages and represent some of the simplest viral entities on earth. Filamentous phages assemble into rods from five different structural proteins, the length of which is proportional to the size of the genome. Generally, the proteinaceous helical rod consists of approximately 2700 copies of the major capsid protein with pentamers of pIII and pVI on one end and pVII and pIX on the other (Abedon 2012a; Ackermann 2003).

3.3.3 ssRNA Phages

Single-stranded RNA phages from the *Leviviridae* family include phages MS2, f2 and ΦCb5. In this morphological group, the highly specific *Leviviridae* capsid is characterized by ninety dimers of the capsid protein arranged into an icosahedral lattice. The virions also assemble with a maturation protein on the capsid to mediate phage attachment and the phage genome is packaged inside the capsid upon maturation (Aksyuk and Rossmann 2011).

While phages have been classified according to morphology it is important to note that inter- and intra-genic modules of information can be combined to perpetually generate new "species" of phage due to co-infection and recombination. As such, phages that share structural similarity may be comprised of vastly different genetic systems for propagation and sustainability.

4 Hurdles for Phage-Based Therapeutics

The small-size, genetic malleability and ease of production of bacteriophages make them ideal candidates for many biotechnological applications. However, no therapeutic has ever been produced without a limitation(s) and phage are no exception. Perhaps one of the major obstacles facing the use of phages for clinical applications is the perception of viruses to the public as "enemies of life" thus imparting a lack of enthusiasm towards phage-based therapeutics (Merabishvili et al. 2009). This issue is further complicated by the documented previous failures in phage-based therapeutics, where phages were used unsuccessfully as antimicrobial agents—an outcome related much more to the lack of understanding of the phage themselves rather than the potential of the technology.

Two critical points in the use of phage-based therapeutics are necessary to address in order to make a substantial impact in the field: (1) improving passaging capacity to create long-circulating phage that can evade the mammalian immune systems; and (2) generating efficiencies in phage scale-up and manufacturing processes.

Phages are quickly removed from a mammalian host by the reticuloendothelial system (RES), a part of the innate mammalian immune system (Lu and Koeris 2011).

New drug delivery technologies, including polymer-based coatings have been shown to enhance phage uptake and reduce phage inactivation/clearing by the RES (Goodridge 2010; Lu and Koeris 2011). In vitro/in vivo evolution of phages could also be used to produce nanoparticles with enhanced properties, including decreased clearance by the host immune system (Merril et al. 1996; *see Chap. 7 for further discussion on phage immune responses and immunomodulation*).

Issues with phage immunity are further complicated by the phage manufacturing and isolation processes. While phage manufacturing has reached a sophistication level worthy of clinical grade products (Merabishvili et al. 2009; Strój et al. 1999; Tanji et al. 2004), isolating phages from their bacterial hosts is convoluted by the presence of endotoxins and pyrogens that are released during phage-induced lysis (Lu and Koeris 2011). As such, there is currently a dearth of well defined and safe manufacturing protocols (Merabishvili et al. 2009) to form safe and stable formulations (Lu and Koeris 2011). Merabishvili et al. (2009) were the first to successfully demonstrate a small-scale, laboratory-based production and application of bacteriophage cocktails system to overcome some of the prevailing issues associated with the efficiency of phage isolation and purification—most notably, the use of a commercially available endotoxin removal kit able to attain efficient purity needed for a European clinical trial (Merabishvili et al. 2009). This group, among others, have addressed these issues (Górski et al. 2005; Yacoby and Benhar 2008), though standard manufacturing procedures are still in demand and are required before phage-based therapeutics can be marketable as legitimate biologics.

References

Abedon, S. T. (2012a). Phages. In P. Hyman & S. T. Abedon (Eds.), *Bacteriophages in health and disease* (pp. 1–5). London: Advances in Molecular and Cellular Microbiology.

Abedon, S. T. (2012b). Phages. In S. T. Abedon & P. Hyman (Eds.), *Bacteriophages in health and disease* (pp. 1–6). London: Advances in Molecular and Cellular Microbiology.

Abedon, S. T., & Thomas-Abedon, C. (2010). Phage therapy pharmacology. *Current Pharmaceutical Biotechnology, 11*(1), 28–47.

Ackermann, H. W. (2003). Bacteriophage observations and evolution. *Research in Microbiology, 154*, 245–251.

Ackermann, H. W. (2007). 5500 Phages examined in the electron microscope. *Archives of Virology, 152*, 227–243.

Aksyuk, A. A., & Rossmann, M. G. (2011). Bacteriophage assembly. *Viruses, 3*(3), 172–203.

Campbell, A. (2003). The future of bacteriophage biology. *Nature Reviews Genetics, 4*(6), 471–477.

Citorik, R. J., Mimee, M., & Lu, T. K. (2014). Bacteriophage-based synthetic biology for the study of infectious diseases. *Current Opinion in Microbiology, 19C*, 59–69.

d'Herelle, F. (1917). Sur un microbe invisible antagoniste des bacilles dysenteriques. *Les Comptes Rendus del'Académie Des Sciences, 165*, 373–375.

Gaidelyte, A., Cvirkaite-Krupovic, V., Daugelavicius, R., Bamford, J. K. H., & Bamford, D. H. (2006). The entry mechanism of membrane-containing phage Bam35 infecting bacillus thuringiensis. *Journal of Bacteriology, 188*(16), 5925–5934.

Goodridge, L. D. (2010). Designing phage therapeutics. *Current Pharmaceutical Biotechnology, 11*(1), 15–27.

Górski, A., Kniotek, M., Perkowska-Ptasińska, A., Mróz, A., Przerwa, A., Gorczyca, W., Nowaczyk, M., et al. (2005). Bacteriophages and transplantation tolerance. *Transplantation Proceedings, 38*(1), 331–333.

Hanlon, G. W. (2007). Bacteriophages: An appraisal of their role in the treatment of bacterial infections. *International Journal of Antimicrobial Agents, 30*(2), 118–128.

Henry, M., & Debarbieux, L. (2012). Tools from viruses: Bacteriophage successes and beyond. *Virology, 434*(2), 151–161.

Jiang, F., & Doudna, J. A. (2015). The structural biology of CRISPR-Cas systems. *Current Opinion in Structural Biology, 30*, 100–111.

Kropinski, A. M. (2006). Phage therapy—Everything old is new again. *Ammi Canada Annual Meeting Symposium, 17*(5), 297–306.

Labrie, S. J., Samson, J. E., & Moineau, S. (2010). Bacteriophage resistance mechanisms. *Nature Reviews Microbiology, 8*(5), 317–327.

Lu, T. K., & Koeris, M. S. (2011). The next generation of bacteriophage therapy. *Current Opinion in Microbiology, 14*(5), 524–531. doi:10.1016/j.mib.2011.07.028

Mardanov, A. V., & Ravin, N. V. (2007). The antirepressor needed for induction of linear plasmid-prophage N15 belongs to the SOS regulon. *Journal of Bacteriology, 189*(17), 6333–6338.

Mcnerney, R., Kambashi, B. S., Kinkese, J., Tembwe, R., & Godfrey-faussett, P. (2004). Development of a bacteriophage phage replication assay for diagnosis of pulmonary tuberculosis. *Society, 42*(5), 2115–2120.

Merabishvili, M., Pirnay, J. P., Verbeken, G., Chanishvili, N., Tediashvili, M., Lashkhi, N., Vaneechoutte, M., et al. (2009). Quality-controlled small-scale production of a well-defined bacteriophage cocktail for use in human clinical trials. *PLoS ONE, 4*(3).

Merril, C. R., Biswas, B., Carlton, R., Jensen, N. C., Creed, G. J., Zullo, S., et al. (1996). Long-circulating bacteriophage as antibacterial agents. *Proceedings of the National Academy of Sciences of the United States of America, 93*(8), 3188–3192.

Morita, M., Tanji, Y., Mizoguchi, K., Akitsu, T., Kijima, N., & Unno, H. (2002). Characterization of a virulent bacteriophage specific for *Escherichia coli* O157:H7 and analysis of its cellular receptor and two tail fiber genes. *FEMS Microbiology Letters, 211*(1), 77–83.

Olszowska-Zaremba, N., Borysowski, J., Dabrowska, K., Górski, A., Hyman, P., & Abedon, S. T. (2012). Phage translocation, safety and immunomodulation. In P. Hyman & S. T. Abedon (Eds.), *Bacteriophages in health and disease* (pp. 168–184).

Randall-Hazelbauer, L., & Schwartz, M. (1973). Isolation of the bacteriophage lambda receptor from Escherichia coli. *Journal of Bacteriology, 116*(3), 1436–1446.

Roberts, J. W., & Devoret, R. (1983). Lysogenic induction. In R. W. Hendrix, J. W. Roberts, F. W. Stahl, & R. A. Weisberg (Eds.), *Lambda II* (pp. 123–144). Cold Springs Harbor, New York: Cold Spring Harbor Laboratory.

Samson, J. E., Magadán, A. H., Sabri, M., & Moineau, S. (2013). Revenge of the phages: Defeating bacterial defences. *Nature Reviews Microbiology, 11*(10), 675–687.

Strój, L., Weber-Dabrowska, B., Partyka, K., Mulczyk, M., & Wójcik, M. (1999). Successful treatment with bacteriophage in purulent cerebrospinal meningitis in a newborn. *Neurologia i Neurochirurgia Polska, 33*(3), 693–698.

Summers, W. C. (1999). *Felix d'Herelle and the origins of molecular biology*. Yale University Press.

Summers, W. C. (2001). Bacteriophage therapy. *Annal Review of Microbiology, 55*, 437–451.

Tanji, Y., Shimada, T., Yoichi, M., Miyanaga, K., Hori, K., & Unno, H. (2004). Toward rational control of *Escherichia coli* O157:H7 by a phage cocktail. *Applied Microbiology and Biotechnology, 64*(2), 270–274.

Twort, F. W. (1915, December 4). An investigation on the nature of ultra-microscopic viruses. *The Lancet.*

Valyasevi, R., Sandine, W. E., & Geller, B. L. (1990). The bacteriophage KH receptor of *Lactococcus lactis* subsp. Cremoris KH is the rhamnose of the extracellular wall polysaccharide. *Applied and Environmental Microbiology, 56*(6), 1882–1889.

Yacoby, I., & Benhar, I. (2008). Targeted filamentous bacteriophages as therapeutic agents. *Expert opinion on drug delivery, 5*(September), 321–329.

Young, R. (2014). Phage lysis: Three steps, three choices, one outcome. *Journal of Microbiology, 52*(3), 243–258.

Chapter 2
Phage for Biocontrol

Abstract Bacteriophage (phage) therapy, or the therapeutic use of phage for the treatment of bacterial diseases, is a classical approach that was originally disregarded due to inconsistent results and with the advent of antibiotic drugs. However, with a greater understanding of phage biology and the pressing need for new and innovative antimicrobial strategies to challenge the ever-increasing prevalence of multidrug-resistant bacterial pathogens, phage therapy is seen to have great potential for reintroduction as antimicrobial strategy, although not without many limitations. In this chapter, by pointing out the limitations of native bacteriophage (phage) therapy, engineered phage-based bactericidal delivery vehicles will be introduced as a treatment approach for the biocontrol of a variety of important pathogens. Such an efficient approach would be suitable for concurrent treatment with standard antibiotics and possibly become a suitable replacement. The bacterial infections to be considered will include those due to: *Escherichia coli*, *Staphylococcus aureus*, *Chlamydia trachomatis*, *Pseudomonas aeruginosa*, and *Helicobacter pylori*. The pathogens will be described along with the efficiency of the phage-based methods to be investigated.

1 Introduction

One of the most concerning problems in therapeutic medicine today is the emergence of multi-drug resistant bacteria and fungi (Sulakvelidze et al. 2011). Bacterial infections are among the most prevalent causes of illness and mortality in clinical settings (Georgiev 2009). The increase of immunosuppressed patients in the present era results in more serious diseases and prolonged hospitalizations with bacterial pathogens (Sulakvelidze et al. 2011; Lu and Collins 2009). Moreover, new antibiotics are not being produced at a sufficient rate to replace the previous medicines which are less effective (Coates and Hu 2007; Kutateladze and Adamia 2010). The economic burden of antibiotic resistance is continuously increasing and

© The Author(s) 2016

J. Nicastro et al., *Bacteriophage Applications—Historical Perspective and Future Potential*, SpringerBriefs in Biochemistry and Molecular Biology, DOI 10.1007/978-3-319-45791-8_2

is currently exceeding an estimated 55 billion dollars annually in the United States alone (Smith and Coast 2013). Additionally, the potential cost of the future development of drug resistance is still unknown. Therefore, adequate attention and the devotion of resources devoted towards resolving the problem of antibiotic-resistant bacteria is one of the first priorities in modern medicine (Sulakvelidze et al. 2011).

Bacteriophages are among the most well studied and abundant organisms on the planet (Clokie et al. 2011). They are distinguished as viral entities that exclusively infect bacterial cells and are composed of a DNA or an RNA genome surrounded by a protein coat. There are two typical phage growth cycles: lytic and non-lytic (Petty et al. 2007), both of which use the host bacterium as a source for their own replication (for a full description about phage types and cycles, refer to Chap. 1: Phage Basics). Considering that phage have a natural capacity to target, exponentially replicate within, and kill their bacterial hosts, they have been deemed a considerable potential option in treatment of bacterial infections (Merril et al. 2003).

Phage therapy can be defined as the therapeutic use of bacteriophage to cure bacterial infections. The history of the usage of bacteriophage therapy for bacterial infections in humans is extensive and dates back to initial studies in this field, as early as 1919 when the co-discoverer of the bacteriophage Felix d'Herelle suggested the use of phage for the treatment of bacterial-induced diarrhoea (Brüssow 2005). Phage-based therapies were sold by American pharmaceutical companies in the 1930s and were used by soldiers in the Second World War to fight off dysentery (Brüssow 2005). The use of phage therapy in the West was thwarted by the invention and practical application of antibiotics for the treatment of bacterial infections (Matsuzaki et al. 2005). However, phage therapy continues to be a common treatment method in the Soviet Union where a number of companies, namely Microgen Inc., sell a long list of different phage cocktails due to a shortage of antibiotics (Hagens et al. 2004; Alisky et al. 1998; Hanlon 2007). Recently, the increasing rate of emergence of multi-drug-resistant bacteria has motivated medical scientists to reconsider phage therapy as a therapeutic option for bacterial infections that are not treatable by conventional antibiotic therapy (Matsuzaki et al. 2005). Even though it is unlikely that antibiotics will be replaced by phages in near future, they offer a great alternative for treatment of drug-resistant pathogens either as monotherapy or in combination with other antibiotics (Kutateladze and Adamia 2010; Smith and Huggins 1982; Kutter et al. 2010).

As a result of problems encountered in using the native phages for treatment of infectious disease, scientists have recently presented the idea of creating genetically modified phages with high killing efficiencies (Hagens et al. 2004). In this chapter we will describe using genetically modified bacteriophage, herein referred to as recombinant phage, for the treatment of bacterial infections. Furthermore, new applications using engineered phages for treatment of drug addictions such as that for cocaine will be briefly discussed.

2 The Importance of Using Recombinant Phage

The past use of native phages for the treatment of bacterial pathogens has not been without its difficulties and consequences, and the stigma arising from these difficulties has led to a false understanding about the potential of phage-based therapeutics (Brüssow 2012). With our current understanding of phage properties and genetics, the limitations of phage therapy using lytic phages can be circumvented with the use of recombinant phages, with their distinct set of properties, for the effective treatment of bacterial diseases. In this section, some of the limitations of native phages will be discussed as well as the alternatives that recombinant phage can offer.

Lysis of bacterial cells, normally associated with lytic phage, will result in the disintegration of cell wall components and consequently the release of endotoxin, typically resulting in inflammation and seen as circulatory shock or sepsis in treated subjects (Paul et al. 2011; Matsuda et al. 2005). To address this limitation, delivery agents for lethal cargoes have been designed using phage-based in vivo packaging systems to create a lysis-deficient phage and/or non-replicative phage that will have bactericidal activity without destroying the cell wall (Goodridge 2010). This system benefits from being able to selectively kill the target cell without releasing the cell contents, which could potentially cause sepsis or release the intracellular toxin that has been delivered. Methods for developing lethal delivery agents may be based on the elimination of the lysis genes from otherwise lytic phage or may use phages that are intrinsically lysis-deficient (Goodridge 2010).

Hagens and Blasi (2003) evaluated a recombinant M13 filamentous phage encoding lethal proteins for killing bacteria without host-cell lysis. Bacterial survival was determined after infection of a growing *Escherichia coli* culture with bacteriophage M13 that encoded either the restriction endonuclease BglII (phage M13R) gene or two modified phage λ S holin genes. Infection of bacteria with either of the recombinant phage led to a high killing efficiency, notably 99.9 % of the host cells were killed within 6 h after treatment with the phage expressing restriction endonuclease BglII (Hagens and Blasi 2003). Furthermore, all treatments succeeded in leaving the host cells intact. Bacterial growth did however resume between 2 and 3 h following infection due to the emergence of phage-resistant mutants (Hagens and Blasi 2003).

In another study by Hagens et al. (2004) engineered non-replicating, non-lytic phages were used to treat *Pseudomonas aeruginosa*. The modified phage killed the bacterial pathogen with minimal release of the host endotoxin (Hagens et al. 2004). It also has been found that modified lysis-deficient *Staphylococcal* phages are efficient in killing of methicillin-resistant *S. aureus* without inducing lysis (Paul et al. 2011). In another study, Matsuda et al. (2005) used a modified *E. coli* phage (*t* amber A3 T4) that was genetically altered to contain a mutation in the holin gene, which prevented lysis of the bacterial cells after infection. The phages were able to effectively infect and replicate within the host bacterial cells; however, their inability to lyse the cells prevented liberation of potentially dangerous endotoxins. The bacterial cells were killed, but remained intact. This phage treatment was demonstrated to improve survival using a murine peritonitis model (Matsuda et al. 2005).

The limited bacteriophage host range is another limitation of phage therapy that should be considered. Each phage species will typically have a very limited and specific host range, in that they can only target one species and even in some cases one single strain of bacteria. Therefore, developing modified phages with an expanded host range through the means of synthetic biology is an important priority in the field. Evidence for the value of expanding the phage host range can be seen in the research from Timothy Lu's lab at MIT (Lu and Collins 2009), including an initial study that involved the grafting the gene 3 protein (g3p) of one filamentous bacteriophage (Ike) to another (Fd), thereby extending the filamentous phage host range. (Ike and Fd are two similar filamentous bacteriophages that target their host by attaching to the pili on host surface membranes. Fd infects bacteria bearing F pili, while Ike infects bacteria bearing N or I pili). In this study the recombinant phage was able to infect bacteria bearing either N or F pili (Lu and Collins 2009).

A further challenge is that phage therapy typically results in a bacterial resistance, often within hours in vitro, to the phages. There is an ever-continuing arms race between phage infection and bacterial resistance to phages, where bacteria have established immunity mechanisms as crucial survival phenotypes. These phenotypes include but not limited to: preventing phage absorption (Labrie et al. 2010; Samson et al. 2013), blocking phage DNA entry (Labrie et al. 2010; Samson et al. 2013), restriction-modification systems (Labrie et al. 2010; Samson et al. 2013), the CRISPR/Cas system (Hatoum-Aslan and Marraffini 2014; Deveau et al. 2010), and abortive infection systems (Labrie et al. 2010; Samson et al. 2013; Amati 1961). Therefore, new techniques are needed to reduce the rate of phage resistance. One of these techniques can be to combine the phage with antibiotics (Lu and Koeris 2011). The use of 'phage cocktails' and/or cycling between different phage treatments is another strategy, as well as specifically considering disrupting the phage-resistance mechanisms while designing the engineered phage (Goodridge 2010).

3   Recombinant Phage for the Treatment of Bacterial Infections

The exponential growth of antibiotic resistance has encouraged researchers to find alternative modalities for treatment of bacterial infections. Pathogens showed resistance to penicillin as early as the 1940s and this became clinically significant leading into the 1960s (Alisky et al. 1998). Currently, there are many pathogens that show resistance not only to penicillin but also to third-generation cephalosporin and even vancomycin (Alisky et al. 1998). Lytic phage therapy has been shown to be effective in treatment of drug-resistant pathogens, at least in uncontrolled clinical studies (Brüssow 2012; Goodridge 2010). In this Section, different studies that employ recombinant phages for the treatment of specific bacterial infections will be discussed.

3.1 Escherichia coli (E. coli)

E. coli, a gram-negative bacillus, is considered one of the important health concerns in the Western world. An example of this organism (*E. coli* O157:H7) which is the most common and most studied member of this group was identified as the causative agent of two outbreaks of bloody diarrheal syndrome in 1982 (Griffin and Tauxe 1991; Rangel et al. 2005). *E. coli* and its relatives can cause an impressive range of diseases. In general, the pathogens can be described as gram-negative bacilli, facultative aerobes and members of the Enterobacteriaceae family. They make up a substantial portion of the human colonic flora, and develop there as early as a few hours after birth (Nataro and Kaper 1998). *E. coli* is typically non-pathogenic when confined to the lumen of the gastrointestinal tract; however, certain strains of this species cause human disease when introduced to other areas of the body. There are many different strains, each with a different clinical outcome. Some strains are considered more pathogenic than others, although most are capable of causing disease, especially in immunocompromised hosts. *E. coli* is the predominant culprit in illnesses such as urinary tract infection (UTI) (Karlowsky et al. 2002). Management of these diseases is complicated by drug-resistant infections. Fluoroquinolone and trimethoprim-sulfamethoxasole resistance limit outpatient treatment while cephalosporin resistance limits inpatient treatment (Johnson et al. 2010).

It has been demonstrated that suppressing SOS network (a global response to DNA damage in which the cell cycle is arrested and DNA repair and mutagenesis are induced) in *E. coli* using engineered M13 bacteriophage heightened quinolone efficiency by several orders of magnitude in vitro. SOS network task is repairing the DNA damage (Echols 1981). To disrupt the SOS response, the lexA 3 SOS suppressing gene was inserted into the phage genome. Moreover, treatment of infected mice with modified phage plus the fluoroquinolone antibiotic Ofloxacin, significantly increased their survival compared to unmodified phage plus antibiotic or no phage plus antibiotic. The level of antibiotic-resistant cells was dramatically reduced with the engineered phage. According to this study, the use of phage in combination with antibiotics could decrease antibiotic-resistant mutants that come from the bacterial population exposed to bactericidal agents (Lu and Collins 2009).

Though this is a unique technique for manipulating bacteriophage targeting the SOS network, which is a beneficial pathway in *E. coli*, could weaken the bacteria harboring the phage (Lu and Collins 2009; Citorik et al. 2014). Following this study, Edgar et al. introduced a system using genetically-engineered phage in order to reverse the pathogen's antibiotic resistance (Edgar et al. 2012). *E. coli* mutants resistant to streptomycin due to mutations in the rpsl gene were isolated and transformed with plasmids containing wild type (WT) rpsl. The delivery of WT rpsL gene to the streptomycin-resistant *E. coli* made the mutants significantly more sensitive (approximately a 10-fold increase in bactericidal activity) to this antibiotic. Furthermore, the group was also able to produce an increase in the bactericidal activity of streptomycin on the rpsL mutants through lysogenization with an engineered bacteriophage lambda (λ) strain modified to carry rpsL gene. To establish whether the system is expandable to other antibiotics, phage λ was engineered to contain wild-type copies of gyrA, then lysogenized with nalidixic acid-resistant bacteria. The recombinant phage was able to restore the *E. coli* strain sensitivity to the nalidixic acid antibiotic. According to this study, the proposed system may be practical for treatment of difficult drug resistant bacterial infections (Edgar et al. 2012).

In another study, Westwater et al. (2003) added Gef and ChpBk toxins to the M13 phagemid system to investigate the possibility of using phage as a lethal-agent delivery vehicle. The bacterial loads were reduced by several orders of magnitude both in vitro and in vivo in mice models infected by *E. coli* following the phage-mediated delivery of bactericidal agents. This technology may open new doors in treatment of multi-drug resistant bacterial pathogens (Westwater et al. 2003).

3.2 Staphylococcus aureus (S. aureus)

S. aureus is one of the main causes of hospital- and community-acquired disease (Hiramatsu et al. 2001). The organism has readily developed resistance against therapeutic agents used over the past 50 years. Methicillin-resistant *Staphylococcus aureus* (MRSA) is the most notable example of this phenomenon and was discovered in 1961 (Hiramatsu et al. 2001). MRSA is now a frequent culprit of skin and soft tissue infections in the United States (Klevens et al. 2007). In hospitalized patients, MRSA infections are correlated with longer hospitalization, increased mortality and morbidity, and higher expenses (Klevens et al. 2007). The emergence of multi-drug resistant *S. aureus* has motivated the re-evaluation of phage therapy for this pathogen.

Recently, it has been shown that a recombinant lysis-deficient *S. aureus* phage P954 could rescue immunocompromised mice infected by MRSA without lysing bacterial cells and releasing endotoxin (Paul et al. 2011). Bacteriophage P954 is a temperate phage that was amplified in *S. aureus* strain RN4220. In order to construct the new plasmid, the native endolysin gene was replaced with an endolysin gene disrupted by the chloramphenicol acetyl transferase (CAT) gene. Induction of the parent plasmid with mitomycin resulted in cell lysis while the endolysin-deficient phage P954 did not lyse. The bactericidal activities of parent and recombinant plasmids were comparable and the host range was the same (Paul et al. 2011).

3.3 Chlamydia trachomatis

Chlamydia trachomatis (CT) is an obligate intracellular pathogen which is responsible for genital tract infections in young sexually active women (Somani et al. 2000; Bébéar and de Barbeyrac 2009; Dean et al. 2000). Recently, it has been indicated that chronic asymptomatic chlamydial infections can cause infertility in women (Somani et al. 2000). A high rate of recurrence of chlamydial infections is common in a sexually active population and has been associated with the development of antibiotic-resistant organisms (Bébéar and de Barbeyrac 2009; Somani et al. 2000). The treatment of CT by conventional phage is challenging, because of its intracellular nature. To overcome this problem, Bhattarai et al. (2012) engineered a M13 phage to express integrin-binding peptide Arg-Gly-Asp (RGD), which is a eukaryotic adhesion motif, to facilitate internalization of the phage into the cells. Moreover, CT peptide (polymorphic membrane protein D) was added to RGD-M13 to interfere with CT infection. In this study, the modified phage reduced the CT infection significantly in primary endocervical cells compared to CT infection alone. The engineered M13 phage enhanced cellular internalization and could be considered as a new modality for treatment of CT infection and other sexually transmitted disease (Bhattarai et al. 2012).

3.4 Pseudomonas aeruginosa (PA)

P. aeruginosa (PA) is a common, gram-negative, opportunistic pathogen that is found to be the culprit in many challenging infections in the airways, epithelium and blood systems. As PA is common in immunocompromised and hospitalized patients, it would be ideal to have a treatment strategy that comes with minimal negative health outcomes to the patient (Hilf et al. 1989; Dzuliashvili et al. 2007).

In one study for treatment of *P. aeruginosa* infection, genetically modified non-replicating, non-lytic phage were produced (Hagens et al. 2004). The PA filamentous phage (Pf3) was armed through recombinant DNA technology with the Bg1II restriction endonuclease gene. The recombinant pf3 phage (Pf3R) was able to significantly reduce PA infection in mice with minimum release of endotoxin, showing good potential for this recombinant phage technology (Hagens et al. 2004). Treatment of infected mice by *P. aeruginosa* with three times the minimal lethal dose (MLD) of either Pf3R or replicating lytic phage resulted in a cure of mice in both cases. In spite of that, when mice were challenged by 5 times the MLD, the survival rate of Pf3R treatment was significantly higher than that of mice treated by lytic phage therapy. This might be due to a reduced inflammatory response in Pf3R-treated mice compared to mice treated by lytic phage. This study demonstrates that treatment of experimental bacterial infection by non-replicative phage can be as effective as replicative phage. Moreover, the use of non-replicative phage would give us the opportunity to specify the therapeutic phage dose, which is not feasible by replicative phage as they increase exponentially (Hagens et al. 2004).

3.5   Helicobacter pylori (H. pylori)

Helicobacter pylori infection is one of the most common pathogens associated with gastritis and both gastric and duodenal ulcers. *H. pylori* has also been connected with mucosa-associated lymphoid tissue (MALT) lymphomas, which have been linked with gastric cancer. Antibiotics currently remain the antibacterial therapeutic choice for *H. pylori* infections; however, there is a need for new and improved strategies (Cao et al. 2000). Cao et al. (2000) have shown that infection by the recombinant ScFv-expressing phage reduced the concentration of all six strains of *H. pylori* in vitro. Moreover, phage treatment of mice infected with *H. pylori* results in a significant decrease in bacterial colonization in the gastric mucosa. To produce this phage, *H. pylori*-antigen-single-chain variable fragments were extracted from murine hybridomas secreting monoclonal antibodies and then expressed as a fusion protein on a filamentous M13 phage (Cao et al. 2000). According to this data, engineered bacteriophages have a good potential in treatment of *H. pylori* and other bacterial pathogens.

4   Phage as Drug Delivery Vehicles for the Treatment of Bacterial Infections

In nanobiotechnology, bacteriophages have been exploited as the gene-delivery cargo for the transfer of gene into mammalian cells since the original identification of internalized phages from libraries of phage-displayed peptide. Recent studies have demonstrated that phage can be a good potential carrier of cytotoxic drugs to apply against both cancer cells and bacterial infections (Yacoby and Benhar 2008; Bar et al. 2008). In one study, filamentous bacteriophages were genetically modified to display P8 coat protein molecule on their surface while chloramphenicol was attached to the bacteriophage through chemical conjugation. Then, the phages were targeted to attach to *S. aureus* bacteria. The results show a retardation of growth of *S. aureus* following treatment with the chloramphenicol-conjugated *S. aureus* targeted phages in comparison to *S. aureus* treated by phages without the cytotoxic drug. In this study the reduction in bacterial growth was not significant because of hydrophobicity of the chloramphenicol, which results in an irreversible precipitation with conjugation of more than 3000 chloramphenicol molecules. To address this limitation, Yacoby and Benhar (2008) applied aminoglycoside antibiotics as a solubility-enhancing linker to connect the hydrophobic drug (i.e. Chloramphenicol) to the phage. The ability of targeted drug-carrying phages to inhibit the growth of methicillin-resistant Staphylococcus, Streptococcus pyogenes, and pathogenic *E. coli* O15787 were tested and complete growth inhibition was obtained (Yacoby and Benhar 2008; Yacoby et al. 2006). To assess the effect of the drug-carrying phages on animals, mice were injected with the recombinant phage. Neomycin-chloramphenicol (Neo-CAM) phages have shown low toxicity in vivo. Moreover,

Neo-CAM carrying phages were less immunogenic in comparison to native unconjugated phage particles (Vaks and Benhar 2011). Targeted drug-delivery may open up new ways in treatment of resistant bacterial pathogens. Furthermore, some potent bactericidal agents are inefficient due to lack of selectivity and this can be solved by targeted therapy (Yacoby et al. 2006).

5 Summary

In this chapter the importance of finding new strategies for the treatment of antibiotic-resistant bacteria as a first priority in modern medicine was emphasized. Native phage therapy was introduced as one of well-known approaches and its limitations were discussed. To overcome the limitations of phage therapy and make it more efficient than other approaches, the genetically-modified phage was introduced and the results of research on different bacterial infections was presented; including but not limited to, *E. coli*, *S. aureus*, *C. trachomatis*, *P. aeruginosa* (PA), and *H. pylori*.

References

Alisky, J., Iczkowski, K., Rapoport, A., & Troitsky, N. (1998). Bacteriophages show promise as antimicrobial agents. *The Journal of Infection, 36*(1), 5–15.

Amati, P. (1961). Abortive Infection of Pseudomonas aeruginosa and *Serratia marcescens* with Coliphage P1. *Journal of Bacteriology, 83*(2), 433–434.

Bar, H., Yacoby, I., & Benhar, I. (2008). Killing cancer cells by targeted drug-carrying phage nanomedicines. *BMC Biotechnology, 8*(37).

Bébéar, C., & de Barbeyrac, B. (2009). Genital Chlamydia Trachomatis Infections. *Clinical microbiology and infection: The official publication of the European society of clinical microbiology and infectious diseases, 15*, 4–10.

Bhattarai, S. R., Yoo, S. Y., Lee, S. W., & Dean, D. (2012). Engineered phage-based therapeutic materials inhibit *Chlamydia Trachomatis* intracellular infection. *Biomaterials, 33*(20), 5166–5174 (Elsevier Ltd).

Brüssow, H. (2005). Phage therapy: The *Escherichia Coli* experience. *Microbiology (Reading, England), 151*(Pt 7), 2133–2140. doi:10.1099/mic.0.27849-0

Brüssow, H. (2012). What is needed for phage therapy to become a reality in Western medicine? *Virology, 434*, 138–142.

Cao, J., Sun, Yi Qian, Berglindh, T., Mellgård, B., Li, Z Qi, Mårdh, B., et al. (2000). *Helicobacter pylori*-antigen-binding fragments expressed on the filamentous M13 phage prevent bacterial growth. *Biochimica et Biophysica Acta - General Subjects, 1474*, 107–113.

Citorik, R. J., Mimee, M., & Lu, T. K. (2014). Bacteriophage-based synthetic biology for the study of infectious diseases. *Current Opinion in Microbiology, 19C*, 59–69 (Elsevier Ltd).

Clokie, M. R. J., et al. (2011). Phages in nature. *Bacteriophage, 1*(1), 31–45.

Coates, A. R. M., & Hu, Y. (2007). Novel Approaches to developing new antibiotics for bacterial infections. *British Journal of Pharmacology, 152*(8), 1147–1154.

Dean, D., Suchland, R. J., & Stamm, W. E. (2000). Evidence for long-term cervical persistence of *Chlamydia Trachomatis* by omp1 Genotyping. *The Journal of Infectious Diseases, 182,* 909–916.

Deveau, H., Garneau, J. E., & Moineau, S. (2010). CRISPR/Cas system and its role in phage-bacteria interactions. *Annual Review of Microbiology, 64*(January), 475–493.

Dzuliashvili, M, K., Gabitashvili, A., Golidjashvili, Hoile, N., & Gachechiladze, K. (2007). Study of therapeutic potential of the experimental *pseudomonas bacteriophage* preparation. *Georgian Medical News, 6*(6).

Echols, H. (1981). SOS functions, cancer and inducible evolution. *Cell, 25*(1), 1–2.

Edgar, R., Friedman, N., Molshanski-Mor, S., & Qimron, U. (2012). Reversing bacterial resistance to antibiotics by phage-mediated delivery of dominant sensitive genes. *Applied and Environmental Microbiology, 78*(3), 744–751.

Georgiev, V. S. (2009). *National Institute of Allergy and infectious diseases, NIH: Volume 2: Impact on global health.* Springer Science & Business Media.

Goodridge, L. D. (2010). Designing phage therapeutics. *Current Pharmaceutical Biotechnology, 11*(1), 15–27.

Griffin, P. M., & Tauxe, R. V. (1991). The epidemiology of infections caused by *Escherichia Coli* O157:H7, other enterohemorrhagic E. Coli, and the associated hemolytic uremic syndrome. *Epidemiologic Reviews, 13,* 60–98.

Hagens, S., & Blasi, U. (2003). Genetically modified filamentous phage as bactericidal agents: A pilot study. *Letters in Applied Microbiology, 37*(4), 318–323.

Hagens, S., Von Ahsen, U., & Von Gabain, A. (2004). Therapy of experimental *Pseudomonas* infections with a nonreplicating genetically modified phage. *Antimicrobial Agents and Chemotherapy, 48*(10), 3817–3822.

Hanlon, G. W. (2007). Bacteriophages: An appraisal of their role in the treatment of bacterial infections. *International Journal of Antimicrobial Agents, 30*(2), 118–128.

Hatoum-Aslan, A., & Marraffini, L. A. (2014). Impact of CRISPR immunity on the emergence and virulence of bacterial pathogens. *Current Opinion in Microbiology, 17,* 82–90 (Elsevier Ltd).

Hilf, M., Yu, V. L., Sharp, J., Zuravleff, J. J., Korvick, J. A., & Muder, R. R. (1989). Antibiotic therapy for *Pseudomonas aeruginosa* bacteremia: Outcome correlations in a prospective study of 200 patients. *The American Journal of Medicine, 87*(5), 540–546.

Hiramatsu, K., Cui, L., Kuroda, M., & Ito, T. (2001). The emergence and evolution of methicillin-resistant staphylococcus aureus. *Trends in Microbiology, 9*(10), 486–493.

Johnson, J. R., Johnston, B., Clabots, C., Kuskowski, M. A., & Castanheira, M. (2010). *Escherichia coli* sequence type ST131 as the major cause of serious multidrug-resistant *E. Coli* infections in the United States. *Clinical Infectious Diseases : An Official Publication of the Infectious Diseases Society of America, 51*(3), 286–94.

Karlowsky, J. A., et al. (2002). Trends in antimicrobial resistance among urinary tract infection isolates of *Escherichia coli* from female outpatients in the United States. *Antimicrobial agents and chemotherapy, 46*(8), 2540–2545.

Klevens, R. Monina, et al. (2007). Invasive methicillin-resistant *Staphylococcus aureus* infections in the United States. *JAMA, 298*(15), 1763–1771.

Kutateladze, M., & Adamia, R. (2010). Bacteriophages as potential new therapeutics to replace or supplement antibiotics. *Trends in Biotechnology, 28*(12), 591–595 (Elsevier Ltd).

Kutter, E., De Vos, D., Gvasalia, G., Alavidze, Z., Gogokhia, L., Kuhl, S., et al. (2010). Phage therapy in clinical practice: Treatment of human infections. *Current Pharmaceutical Biotechnology, 11*(1), 69–86.

Labrie, S. J., Samson, J. E., & Moineau, S. (2010). Bacteriophage resistance mechanisms. *Nature Reviews Microbiology, 8*(5), 317–327.

Lu, T. K., & Collins, J. J. (2009). Engineered bacteriophage targeting gene networks as adjuvants for antibiotic therapy. *Proceedings of the National Academy of Sciences of the United States of America, 106*(12), 4629–4634.

Lu, T. K., & Koeris, M. S. (2011). The next generation of bacteriophage therapy. *Current Opinion in Microbiology, 14*(5), 524–531 (Elsevier Ltd).

Matsuda, T, Freeman, T. A., Hilbert, D. W., Duff, M., Fuortes, M., Stapleton, P. P., & Daly, J. M., et al. (2005). Lysis-deficient bacteriophage therapy decreases endotoxin and inflammatory mediator release and improves survival in a murine peritonitis model. *Surgery, 137*(6), 639–46.

Matsuzaki, S., Rashel, M., Uchiyama, J., Sakurai, S., Ujihara, T., Kuroda, M., et al. (2005). Bacteriophage therapy: A revitalized therapy against bacterial infectious diseases. *Journal of Infection and Chemotherapy: Official Journal of the Japan Society of Chemotherapy, 11*(5), 211–219.

Merril, C. R., Scholl, D., & Adhya, S. L. (2003) The prospect for bacteriophage therapy in Western medicine. *Nature Reviews Drug Discovery, 2*, 489–497.

Nataro, J. P., & Kaper, J. B. (1998). Diarrheagenic escherichia coli. *Clinical Microbiology Reviews, 11*(1), 142–201.

Paul, V. D., Sundarrajan, S., Rajagopalan, S. S., Hariharan, S., Kempashanaiah, N., Padmanabhan, S., & Sriram, B., et al. (2011). Lysis-deficient phages as novel therapeutic agents for controlling bacterial infection. *BMC Microbiology, 11*(1), 195 (BioMed Central Ltd).

Petty, N. K., Evans, T. J., Fineran, P. C., & Salmond, G. P. C. (2007). Biotechnological exploitation of bacteriophage research. *Trends in Biotechnology, 25*(1), 7–15.

Rangel, J. M., Sparling, P. H., Crowe, C., Griffin, P. M., & Swerdlow, D. L. (2005). Epidemiology of *Escherichia Coli* O157:H7 outbreaks, United States, 1982–2002. *Emerging Infectious Diseases,* 11(4).

Samson, J. E, Magadán, A. H., Sabri, M., & Moineau, S. (2013). Revenge of the phages: Defeating bacterial defences. *Nature Reviews. Microbiology, 11*(10). Nature Publishing Group: 675–87.

Smith, H. W., & Huggins, M. B. (1982). Successful treatment of experimental *Escherchia coli* infections in mice using phage: Its general superiority over antibiotics. *Journal of General Microbiology, 128*, 307–318.

Smith, R., & Coast, J. (2013). The true cost of antimicrobial resistance. *Bmj-British Medical Journal, 346*(March), 5.

Somani, J, Bhullar, V. A., Workowski, K. A., Farshy, C. E., & Black, C. M. (2000). Multiple drug-resistant *Chlamydia trachomatis* associated with clinical treatment failure. *The Journal of Infectious Diseases 181*, 1421–27.

Sulakvelidze, A., Alavidze, Z., Glenn, J., & Morris, J. G. (2001). Bacteriophage therapy. *Antimicrobial Agents and Chemotherapy, 45*(3), 649–59.

Vaks, L., & Benhar, I. (2011). In vivo characteristics of targeted drug-carrying filamentous bacteriophage nanomedicines. *Journal of Nanobiotechnology, 9*(1), 58 (BioMed Central Ltd).

Westwater, C., Kasman, L. M., Schofield, D. A., Werner, P. A., Dolan, J. W., Schmidt, M. G., et al. (2003). Use of genetically engineered phage to deliver antimicrobial agents to bacteria: An alternative therapy for treatment of bacterial infections. *Antimicrobial Agents and Chemotherapy, 47*(4), 1301–1307.

Yacoby, I., & Benhar, I. (2008). Targeted filamentous bacteriophages as therapeutic agents. *Expert opinion on drug delivery, 5*(September), 321–329.

Yacoby, I., Shamis, M., Bar, H., Shabat, D., & Benhar, I. (2006). Targeting antibacterial agents by using drug-carrying filamentous bacteriophages. *Antimicrobial Agents and Chemotherapy, 50*(6), 2087–2097.

Chapter 3
Phage Device Coatings

Abstract The presence of biofilms and their associated antimicrobial resistance provides a challenge to various industries where new and affective device coating strategies are required. Bacteriophages have the natural capacity to act as antibacterials and have been used extensively for this purpose, including in device coatings, since the beginning of the 20th century. This Chapter explores the emerging industry of phage-coated medical devices. An extensive review on the biology and challenges behind biofilm formations, including the contributors to biofilm resistance and current antimicrobial strategies will be covered. Alternative medical device coating strategies will also be explored, including the benefits, challenges, and promise of phage-based device coatings as bioactive agents.

1 Introduction

Biofilms exist throughout the environment as dynamic, viable, and often metabolically-active microbial structures consisting of microbial cells embedded in a matrix of extracellular materials (Azeredo and Sutherland 2008). The composition of the microbial biofilm allows for the cultivation of microorganisms in an array of environments (Sutherland et al. 2004). The structure of a microbial biofilm is dependent on several factors; the particular content of microorganisms, their physiological state and the physical environment (Sutherland 2001; Sutherland et al. 2004). More than three quarters (75–89 %) of extracellular polymeric substance (EPS) is composed of proteins and polysaccharides containing various functional groups such as carboxyls, amino acids, and phosphate groups (Tsuneda et al. 2003) EPS-rich biofilms have also been found to contain relatively high levels of hexose, hexosamine, and ketose groups (Tsuneda et al. 2003). Various researchers have found that some microorganisms complement one another well, having a positive synergistic effect on biofilm formation (Burmølle et al. 2006). It has also been suggested that quorum sensing may have a role in the biofilm's antimicrobial resistance (von Eiff et al. 2005).

© The Author(s) 2016   21
J. Nicastro et al., *Bacteriophage Applications—Historical Perspective*
and Future Potential, SpringerBriefs in Biochemistry and Molecular Biology,
DOI 10.1007/978-3-319-45791-8_3

Bacterial adherence and biofilm formation on vital medical devices such as central venous catheters (CVCs) can lead to persistent infections with a mortality rate of 12–25 % for each CVC bloodstream infection (Stewart and Costerton 2001; von Eiff et al. 2005). As reported in the early 2000s, the annual cost of CVC-associated bloodstream infections (BSIs) in the US may range from $US 296 million to $US 3.2 billion (von Eiff et al. 2005; Mermel 2000) with an estimated $US 25,000 per infection (von Eiff et al. 2005; Safdar et al. 2002). The demand for a safer alternative is clear.

2   Biofilms on Medical Devices

Medical treatment often requires indwelling or subcutaneous biomedical implants, particularly in the hospital setting. Microbial biofilms can form on these implants and cause an infection which is associated with many pathologies (Azeredo and Sutherland 2008), such as endocarditis, urinary tract infections (UTIs), or respiratory infections in cystic fibrosis patients (Burmølle et al. 2006). Indwelling catheters such as CVCs can cause catheter-related infections (CRIs) and result in local infections or BSIs (Curtin and Donlan 2006).

Emerging resistance against current antibiotic therapies, particularly in the hospital setting and with the resultant compromise to host defenses, is a concern and motivation for alternatives (Azeredo and Sutherland 2008). Inaccessibility to biomedical implants that are contaminated combined with poor drug penetration of drug molecules into microbial biofilm structures are contributing challenges to this issue in the healthcare setting (Stewart and Costerton 2001). Medical device contamination most likely occurs during inoculation of a few microorganisms into the patient's skin or mucous membrane during implantation; medical staff may also be carriers to pathogenic organisms (von Eiff et al. 2005).

Multiple factors allow for microbial adherence to foreign material such as that of medical devices. The initial adherence of the microorganism to the medical device is dependent on physiochemical forces such as polarity (von Eiff et al. 2005). Two surface-associated proteins associated with polystyrene biomaterial adhesion for *Staphylococcus epidermis* have been demonstrated (Veenstra et al. 1996), partially assisted by surface-associated autolysin (Heilmann et al. 1997).

Upon infection, the opportunistic microorganisms may rapidly attach to the biomaterial and accumulate in the host body to form multi-layered clusters (von Eiff et al. 2005) with the assistance of biofilm-associated protein (Bap) in *Staphylococcus aureus* (Cucarella et al. 2001). Implanted devices in the body quickly become coated with plasma and connective tissue proteins which subsequently serve as specific receptors for colonizing microorganisms (Mack et al. 1996). Polysaccharide structures called capsular polysaccharide/adhesion have been associated with the initial adherence and slime production (Muller et al. 1993). Host-factor binding proteins from coagulase-negative staphylococci (CoNs) have also been found along with teichoic acid, an element found on the cell walls of the microorganism, which is suggested to function as a bridging molecule between bacteria and fibronectin-coated polymer (Heilmann et al. 1997; Hussain et al. 2001).

In a report published by the Centers for Disease Control and Prevention (CDCP), biofilms commonly found on medical devices such as catheters and mechanical heart valves may possess bacterial flora that originate from species of gram-positive typically found on the patient's skin or exogenous microflora from health-care personnel, however, urinary catheters contain mostly gram-negative microorganisms (Donlan 2001). The primary causative agent for foreign body related infections (FBRIs) associated with medical devices are CoNs, particularly *S. epidermis*, a gram-positive microorganism that typically resides on the human skin and mucous membranes (Hugonnet et al. 2004).

2.1 Contributors to Biofilm Resistance

Many infectious biofilms contain a single bacterial species, however, in certain multispecies biofilms, synergistic reactions between the bacteria have been seen to enhance biofilm formation typically resulting in an increased resistance to antimicrobial agents (Azeredo and Sutherland 2008; Burmølle et al. 2006). One contributor towards biofilm antibiotic resistance is accessibility. Polysaccharides in biofilms have been demonstrated to decrease diffusion of antibiotics and their antimicrobial activity (Allison and Matthews 1992).

In a study conducted by Burmølle et al. (2006), biofilms consisting of more than one strain of bacteria demonstrated higher resistance than single-species biofilm when exposed to antibacterial agents. This was demonstrated for a four-strain biofilm, with increased protection against all the antibacterial and invasive species suggesting increased fitness due to the synergistic effects from excreted compounds. These researchers noted that certain bacterial species in the study had specific roles in the multispecies synergy and that this was dependent on the presence of the other species (Burmølle et al. 2006). One of the antibacterial agents used was hydrogen peroxide, to simulate oxidative stress caused by reactive intermediates in metabolism, from degradation of organic matter, or found naturally produced by other microorganisms such as phagocytes. Tetracycline was also selected because it is an agent typically used in disinfectants and treatment of microbial infections (Burmølle et al. 2006). In addition to chemical treatment, the biofilms were challenged with *Pseudoalteromonas trunicata*, a marine bacterium that produces a range of biocidal compounds, particularly AlpP, that are effective against gram-negative and gram-positive isolates. Expression of AlpP in *P. trunicata* biofilms has been previously demonstrated to play a role in competitive dominance during mixed-species biofilm formation (Rao et al. 2005).

The biofilm formed by the mixed species showed a more viscous matrix, a feature which would decrease antimicrobial permeability and effectiveness, thereby enhancing antibacterial resistance (Allison and Matthews 1992; Burmølle et al. 2006). Following antibacterial exposure, changes in the biofilm matrix were suggested to occur therefore further reducing the permeation of antimicrobial agents. The enhanced biofilm formation from mixed-species may also be due to enzyme complementation between certain species. Single-species biofilms may grow slower or encounter cell death at lower cell densities due to nutrient depletion, in contrast with mixed-species biofilms, assuming that components of the latter are not in direct competition for resources (Burmølle et al. 2006). In a study conducted by Leriche et al. (2003), it was suggested that exposure to antimicrobial agents actually induced cells in biofilm to coexist as mixed structure—suggestive of protection of members in the biofilm by other more resistant species (Leriche et al. 2003).

Quorum sensing is a suggested contributor to antibiotic resistance. When imposing various conditions on *S. aureus*, its quorum sensing accessory gene regulator (*agr*) locus had a varied effect on biofilm formation, however through these studies it was found that the signaling *agr* mutants were particularly sensitive to rifampicin, a bactericidal antibiotic drug, but not oxacillin, a low-spectrum antibiotic of the penicillin class (von Eiff et al. 2005).

3 Alternative Medical Coating Devices

Various alternate preventative measures have been established. One such mechanism that is used is the antibiotic-lock technique for intraluminal therapy (von Eiff et al. 2005). Antibiotic-lock technique is the installation of a highly concentrated antibiotic solution into the intravascular catheter lumen for the purposes of sterilization (von Eiff et al. 2005). Strong microbial killing effects are demonstrated in catheters in vitro using antibiotics, however, studies using *S. aureus* demonstrate that pharmacokinetic parameters do not necessarily correspond to efficacy, particularly with the rise in methicillin-resistant *S. aureus* (MRSA) strains (Curtin and Donlan 2006; Espersen et al. 1994). Alternatively, coating medical devices or their parts with antimicrobial drugs has been practiced with the purpose of these activities bound directly or within an aqueous environment for drug release; but this faces the same risk of bacterial resistance (Solovskij et al. 1983; Sherertz et al. 1993).

The emergence of multi-resistant bacterial strains has triggered a renewed interest for alternative coating methods such as the development of anti-adhesive surfaces (Curtin and Donlan 2006; von Eiff et al. 2005). Using a modified polystyrene surface, several researchers demonstrated a substantial reduction in bacterial adhesion with modified surfactant, of 70–90 % (Bridgett et al. 1992; Desai et al. 1992). However, its anti-adhesion properties were diminished or impaired in the presence of proteins in vivo (Curtin and Donlan 2006).

4 Bacteriophage as Bioactive Coatings

The renewed interest in alternative coating methods for medical devices has turned attention to bacteriophage coating. Phage, which have the natural ability to specifically target and eliminate their bacterial hosts without much damage to mammalian cells, are an attractive alternative to microbial sanitizers which lack specificity and are most often toxic. Lytic phages are particularly effective for medical coating purposes due to their abilities for bacterial cell lysis and their availability in abundance where they are most required. Lytic phage also do not integrate into host DNA (Azeredo and Sutherland 2008).

Phages are not only the natural, specific predators of bacteria, they have also co-evolved with bacteria in such a way that biofilm growth in bacteria would have been compensated for in phage evolution. Evidence for this evolution has been shown in a study by Hibma et al. who bred a *Listeria*-specific phage that could prevent biofilm formation and also remove pre-existing biofilms of L-form *Listeria monocytogenes* on stainless steel in comparable amounts to lactic acid (Hibma et al. 1997). Phages are also capable of traversing through bacterial biofilms either through diffusion, collision or attachment to target bacterial cells (Lu and Collins 2007; Donlan 2009; Curtin and Donlan 2006).

To exploit the natural abilities of phage for the control of biofilms. Barr et al. (2013) proposed the Bacteriophage Adherence to Mucus (BAM) model. By adhering bacteriophage to mucus, this approach provides a ubiquitous but non-host-derived immunity through binding interactions between mucin glyco-proteins and Ig-like protein domains exposed on phage capsids. In the study, Barr et al. utilized the lung mucosal system which has a physical and biochemical antimicrobial defense mechanism itself. Through a phage-mucosal enrichment process, their strategy demonstrated a reduction in microbial colonization and pathology using the model T4 phage in vitro, by four orders of magnitude. The adherence of phage also protected mucus-producing cells against subsequent wave of bacteria, suggesting a sustained antimicrobial response. The presence of mucus for the progeny and bacteriophage adherence proved important since phage pre-treatment on mucus-producing cells was attributed with a 3.6-fold reduction in bacterial growth, compared to the knock-out mutant control that received the same pre-treatment. The authors add that as the system becomes more complex, a dynamic 'arms race' between the phages and the infectious bacteria may occur naturally while protecting the host; thus creating a metazoan-phage symbiosis (Barr et al. 2013).

The adhesion between the mucosal layer and bacteriophage in the Barr et al. (2013) studies exploited the role of capsid Ig-like domains. Four Ig-like domains were found in high antigenic outer capsid protein (Hoc) and it is hypothesized that the 155 copies displayed on the capsid surface of the T4 Hoc might have a role in mediating the adherence of T4 phage to the mucus layer. The presence of the Ig-like protein (Hoc) displayed on the capsid of T4 phage significantly slowed the diffusion of the phage in mucin solutions and mucin glycans thus allowing for increased residence time and potentially increasing their replicative success and subsequent antimicrobial properties (Barr et al. 2013).

An alternative use of bacteriophage for the control of biofilms is by utilizing its ability to degrade biofilm matrices. Research conducted by Carson et al. (2010) focused towards urinary tract infections attributed to indwelling urethral catheters. Causative agents of catheter-associated urinary tract infections (CAUTI) are typically gram-negative in nature, primarily *P. aeruginosa*, although gram-positive *S. epidermis* is commonly isolated (Carson et al. 2010). The primary complication of CAUTI is due to encrusted mineralized deposits which are stabilized by bacterial biofilm leading to blockage and subsequent infection (Azeredo and Sutherland 2008). The issue is further compounded by high urinary pH and the presence of *Proteus mirabilis* swarmer cells making successful treatment extremely difficult (Azeredo and Sutherland 2008). This research group sought to utilize T4 bacteriophages against *Escherichia coli* by combining the ability of phage to degrade biofilm matrices with a hydrogel coating technology, resulting in a 99.9 % reduction in biofilm; the reduction was attributed to the access within the biofilm by bacteriophage depolymerase (Carson et al. 2010).

Similarly, Hanlon et al. (2001) demonstrated that the bacteriophage depolymerase enzyme alone is capable of disrupting biofilm structure viscosity by 40 % (Hanlon et al. 2001). Lu and Collins (2007) have also engineered bacteriophages to express potent EPS-degrading enzymes, which substantially reduced bacterial biofilm counts by more than 10^4-fold—demonstrating the feasibility and ability to develop tailored anti-biofilm therapies for coating (Lu and Collins 2007).

Combining these various approaches to form a bioactive coating may significantly diminish blockage, particularly with urethral catheters when combined with bladder wash-out solutions, thereby mitigating the risk of infection (Carson et al. 2010).

5 Conclusions

The quest for new and promising antimicrobial strategies is needed for the battle against the persistence of clinically-relevant device-associated biofilm infections in the healthcare industry along with the rise in antimicrobial resistance. Bacteriophages have been used for decades for the treatment of human infections, and their natural ability that can be exploited for the treatment of clinically relevant biofilms, particularly when incorporated as device coatings on medical devices for the prevention of biofilms. The nature of device-associated biofilms means that the most effective strategies, and a realm of potential for clinically relevant device coatings, will lie where biofilm formation is prevented. In cases where biofilms currently persist, effective strategies involving phage will target the extracellular matrix component of the biofilm that could act in tandem with antimicrobial agents. Furthermore, phage can be engineered for even greater efficacy against biofilms providing an endless potential for bacteriophage in the field.

References

Allison, D. G., & Matthews, M. J. (1992). Effect of polysaccharide interactions on antibiotic susceptibility of *Pseudomonas aeruginosa*. *The Journal of Applied Bacteriology, 73*(6), 484–488.

Azeredo, J., & Sutherland, I. W. (2008). The use of phages for the removal of infectious biofilms. *Current Pharmaceutical Biotechnology, 9*, 261–266.

Barr, J. J., Auro, R., Furlan, M., Whiteson, K. L., Erb, M. L., Pogliano, J., et al. (2013). Bacteriophage adhering to mucus provide a non-host-derived immunity. *Proceedings of the National Academy of Sciences, 110*(26), 10771–10776.

Bridgett, M. J., Davies, M. C., & Denyer, S. P. (1992). Control of staphylococcal adhesion to polystyrene surfaces by polymer surface modification with surfactants. *Biomaterials, 13*(7), 411–416.

Burmølle, M., Webb, J. S., Rao, D., Hansen, L. H., Sørensen, S. J., & Kjelleberg, S. (2006). Enhanced biofilm formation and increased resistance to antimicrobial agents and bacterial invasion are caused by synergistic interactions in multispecies biofilms. *Applied and Environmental Microbiology, 72*(6), 3916–3923.

Carson, L, Gorman, S. P., & Gilmore, B. F. (2010). The use of lytic bacteriophages in the prevention and eradication of biofilms of proteus mirabilis and *Escherichia coli*. *FEMS Immunology & Medical Microbiology, 59*(3), 447–455.

Cucarella, C., Solano, C., Valle, J., Amorena, B., Lasa, I., & Penadés, J. R. (2001). Bap, a *Staphylococcus aureus* surface protein involved in biofilm formation. *Journal of Bacteriology, 183*(9), 2888–2896.

Curtin, J. J., & Donlan, R. M. (2006). Using bacteriophages to reduce formation of catheter-associated biofilms by staphylococcus *Epidermidis* using bacteriophages to reduce formation of catheter-associated biofilms by *Staphylococcus* epidermidis. *Antimicrobial Agents and Chemotherapy, 50*(4), 1268–1275.

Desai, N. P., Hossainy, S. F., & Hubbell, J. A. (1992). Surface-immobilized polyethylene oxide for bacterial repellence. *Biomaterials, 13*(7), 417–420.

Donlan, R. M. (2001). Biofilms and device-associated infections. *Emerging Infectious Diseases, 7*(2), 277–281.

Donlan, R. M. (2009). Preventing biofilms of clinically relevant organisms using bacteriophage. *Trends in Microbiology, 17*(2), 66–72. http://www.ncbi.nlm.nih.gov/pubmed/19162482

Espersen, F., Frimodt-Møller, N., Corneliussen, L., Riber, U., Rosdahl, V. T., & Skinhøj, P. (1994). Effect of treatment with methicillin and gentamicin in a new experimental mouse model of foreign body infection. *Antimicrobial Agents and Chemotherapy, 38*(9), 2047–2053.

Hanlon, G. W., Denyer, S. P., Olliff, C. J., & Ibrahim, L. J. (2001). Reduction in exopolysaccharide viscosity as an aid to bacteriophage penetration through *Pseudomonas aeruginosa* biofilms. *Applied and Environmental Microbiology, 67*(6), 2746–2753.

Heilmann, C., Hussain, M., Peters, G., & Götz, F. (1997). Evidence for autolysin-mediated primary attachment of staphylococcus *Epidermidis* to a polystyrene surface. *Molecular Microbiology, 24*(5), 1013–1024.

Hibma, A. M., Jassim, S. A. A., & Griffiths, M. W. (1997). Infection and removal of L-forms of listeria monocytogenes with bred bacteriophage. *International Journal of Food Microbiology, 34*(3), 197–207.

Hugonnet, S., Sax, H., Eggimann, P., Chevrolet, J. C., & Pittet, D. (2004). Nosocomial bloodstream infection and clinical sepsis. *Emerging Infectious Diseases, 10*(1), 76–81.

Hussain, M., Heilmann, C., Peters, G., & Herrmann, M. (2001). Teichoic acid enhances adhesion of staphylococcus *Epidermidis* to immobilized fibronectin. *Microbial Pathogenesis, 31*(6), 261–270.

Leriche, V., Briandet, R., & Carpentier, B. (2003). Ecology of mixed biofilms subjected daily to a chlorinated alkaline solution: spatial distribution of bacterial species suggests a protective effect of one species to another. *Environmental Microbiology, 5*(1), 64–71.

Lu, T. K., & Collins, J. J. (2007). Dispersing biofilms with engineered enzymatic bacteriophage. *Proceedings of the National Academy of Sciences of the United States of America, 104*(27), 11197–11202. doi:10.1073/pnas.0704624104.

Mack, D., Fischer, W., Krokotsch, A., Leopold, K., Hartmann, R., Egge, H., et al. (1996). The intercellular adhesin involved in biofilm accumulation of *Staphylococcus* epidermidis is a linear beta-1,6-linked glucosaminoglycan: purification and structural analysis. *Journal of Bacteriology, 178*(1), 175–183.

Mermel, L. A. (2000). Prevention of intravascular catheter-related infections. *Annals of Internal Medicine, 132*(5), 391–402.

Muller, E., Hübner, J., Gutierrez, N., Takeda, S., Goldmann, D. A., & Pier, G. B. (1993). Isolation and characterization of transposon mutants of *Staphylococcus epidermidis* deficient in capsular polysaccharide/adhesin and slime. *Infection and Immunity, 61*(2), 551–558.

Rao, D., Webb, J. S., & Kjelleberg, S. (2005). Competitive interactions in mixed-species biofilms containing the marine bacterium *Pseudoalteromonas tunicata*. *Applied and Environmental Microbiology, 71*(4), 1729–1736.

Safdar, N., Kluger, D. M., & Maki, D. G. (2002). A review of risk factors for catheter-related bloodstream infection caused by percutaneously inserted, noncuffed central venous catheters: Implications for preventive strategies. *Medicine, 81*(6), 466–479.

Sherertz, R. J., Carruth, W. A., Hampton, A. A., Byron, M. P., & Solomon, D. D. (1993). Efficacy of antibiotic-coated catheters in preventing subcutaneous staphylococcus aureus infection in rabbits. *Journal of Infectious Diseases, 167*(1), 98–106.

Solovskij, M. V., Ulbrich, K., & Kopecek, J. (1983). Synthesis of N-(2-hydroxypropyl) methacrylamide copolymers with antimicrobial activity. *Biomaterials, 4*(1), 44–48.

Stewart, P. S., & Costerton, J. W. (2001). Antibiotic resistance of bacteria in biofilms. *Lancet, 358*(9276), 9135–9138.

Sutherland, I. W., Hughes, K. A., Skillman, L. C., & Tait, K. (2004). The interaction of phage and biofilms. *FEMS Microbiology Letters, 232*(1), 1–6.

Sutherland, Ian W. (2001). Biofilm exopolysaccharides: A strong and sticky framework. *Microbiology, 147*(1), 3–9.

Tsuneda, S., Aikawa, H., Hayashi, H., Yuasa, A., & Hirata, A. (2003). Extracellular polymeric substances responsible for bacterial adhesion onto solid surface. *FEMS Microbiology Letters, 223*(2), 287–292.

Veenstra, G. J., Cremers, F. F., van Dijk, H., & Fleer, A. (1996). Ultrastructural organization and regulation of a biomaterial adhesin of *Staphylococcus epidermidis*. *Journal of Bacteriology, 178*(2), 537–541.

von Eiff, C., Jansen, B., Kohnen, W., & Becker, K. (2005). Infections associated with medical devices: Pathogenesis, management and prophylaxis. *Drugs, 65*(2), 179–214.

Chapter 4
Bacteriophages Functionalized for Gene Delivery and the Targeting of Gene Networks

Abstract Bacteriophages (phages) offer many potential and existing applications to biotechnology, including their modification and use as protein/gene carriers. Phages possess many intrinsic physicochemical attributes that make them excellent candidates for use in gene therapy. In this chapter we will explore how phages have been employed in gene delivery as well as their future utility in this exciting medical application.

1 Introduction to Phage Mediated Delivery of Genetic Material

Bacteriophages were among the first entities to be manipulated for modern gene transfer and gene targeting strategies. The small size, relative ease of production, capacity for genomic isolation and manipulation position phage attractively for this pursuit. The bacteriophage genome can be manipulated to incorporate heterologous sequences designed to be expressed in, or otherwise modify, a recipient cell. Gene expression in a recipient cell can be governed either by prokaryotic or eukaryotic genetic systems, making it possible to deliver genetic cargo that can be expressed in any host cell. The natural host specificity of a phage governs its tropism and when manipulated, provides an endless potential of biotechnological applications. Alternatively, phages can and have also been exploited for use as display instruments (Smith and Petrenko 1997). Known as phage display, this is a strategy of conjugating or translationally fusing peptide molecules onto the coat surface of the phage. Phage display of targeting ligands or antibodies can be generated against a range of mammalian cells for which the phage would naturally have no tropism nor capacity for propagation (Nicastro et al. 2014). Under such applications the phage would no longer function as a (bacterial) virus, but rather as an inert, nanoscale particle employed to deliver nucleic acid cargo and enact a specific activity to a targeted

© The Author(s) 2016

J. Nicastro et al., *Bacteriophage Applications—Historical Perspective and Future Potential*, SpringerBriefs in Biochemistry and Molecular Biology, DOI 10.1007/978-3-319-45791-8_4

mammalian cell. In a seemingly endless assortment of activities and targets, the focus of this chapter will be on the specific characteristics of the engineered phage delivery vehicles and their delivery of encapsulated nucleic acids to their targeted cells.

Phage-delivered nucleic acid cargo to a recipient cell can be benign or toxic. Benign treatments, including gene therapy (Larocca and Baird 2001), immunomodulation (Willats 2002), and phage DNA vaccines (Clark et al. 2011) are intended to leave the cell metabolically active so that the host cell can express the DNA cargo provided by the phage. Phages may also be modified to encode toxins (Vilchez and Jacoby 2004) or other genes that are damaging to the cell (Abedon 2009), which can be used as a means of biocontrol in prokaryotic systems or as a means to kill or damage non-bacterial targets such as tumour cells. The expression of the cargo can be temporary, which is the goal in bacterial identification, bio-control and phage mediated DNA vaccine applications. In other cases, such as in gene therapy in eukaryotic systems and gene cloning in prokaryotic systems, the expression of the phage encoded genetic cargo can and is aimed to be long lasting.

2 Bacteriophages as Gene Delivery Vehicles

Most commonly, eukaryotic viral vectors are employed to deliver DNA to eukaryotic cells to correct for a genetic defect or otherwise augment a desired cell phenotype. However, the use of such vectors poses important safety concerns, particularly with respect to the control of inherent virulence and immunogenicity (Clark et al. 2012; Seow and Wood 2009), which can and have previously resulted in mortality (Somia and Verma 2000). An additional challenge is the effective design of viral therapeutics to target the desired organ or tissue and to avoid prior immunity against viral vectors (Nayak and Herzog 2010). As such, there is a need for gene therapy systems that are benign and yet more precise, which has drawn attention to alternate approaches such as bacteriophage-mediated gene delivery.

Phages are stable (Jepson and March 2004), inexpensive to produce (Bakhshinejad et al. 2014), easy to manipulate genotypically and phenotypically (Clark et al. 2011), and can be targeted to their intended cellular targets (Nicastro et al. 2013). In their application to gene delivery, the outer capsid coat proteins of the phage vehicle can simultaneously protect the intended DNA cargo against degradation during delivery (Clark and March 2006; Dunn 1996) and tolerate capsid peptide/protein fusions, making it possible for the vehicle to target intended cells (Clark and March 2006); a cornerstone of successful gene therapeutic design. Phage can be safely administered to mammals as evidenced by the long history of phage therapy against bacterial infection (Abedon et al. 2011) and do, in fact, naturally penetrate mammalian tissue (Dabrowska et al. 2005) without intrinsically infecting mammalian cells. Despite the high penetrance of phage particles, they are also quickly cleared by the reticuloendothelial system (RES), which will negatively impact uptake by target cells (Molenaar et al. 2002). To circumvent clearance, long-circulating phages capable of evading the RES have been developed

(Merril et al. 1996). Alternatively, the addition of polyethylene glycol (PEG) can also improve phage circulation (Kim et al. 2008).

While the simplicity of phages positions them as attractive cloning vectors, they are generally limited by the size of the gene(s) of interest that can be cloned into the phage head, and in most cases lack natural nuclear honing and expression abilities in eukaryotic cells (Clark et al. 2011; Larocca and Baird 2001). While filamentous bacteriophage have a far more flexible packaging minimum and maximum (Specthrie et al. 1992), they are instead limited by the fact that they must carry circular single-stranded DNA, a detriment to expression in eukaryotic hosts where conversion to double stranded DNA is required for gene expression (Clark et al. 2012; Yacoby and Benhar 2008). Some of these limitations can be addressed by the modification of the phage particle including: the display of cell-penetrating peptides (Trabulo et al. 2012) and/or the use of chemical agents such as DEAE-dextran (Yokoyamakobayashi and Kato 1993), co-administration with cationic lipids (Eguchi et al. 2001; Yokoyamakobayashi and Kato 1994), and the inclusion of nuclear localization peptides and/or sequences (Lam and Dean 2010; Miller and Dean 2009). Despite carrying single-stranded DNA, filamentous phage gene delivery has resulted in successful uptake and expression in mammalian cells (Larocca et al. 1999; Poul and Marks 1999). Additionally, the introduction of self-complementary sequences in filamentous phage single-stranded DNA can lead to the formation of double-stranded DNA in eukaryotic cells (Prieto and Sánchez 2007).

Despite the above limitations, phages have validated their utility as functional gene delivery agents, with the first reported use for this application more than two decades ago by Hart et al. (1994). In this study, phage fd was employed to display a cyclic-binding peptide to the major coat protein pVIII. This amino (N)-terminal fusion, occurring at approximately 300 copies/phage particle, was bound to cells and was efficiently internalized. The same peptide sequence was then fused to the major tail protein (gpV) of bacteriophage λ. These modified phages proved to be more suitable gene delivery candidates—transfecting mammalian cells at a remarkable frequency in comparison to the undecorated controls (Dunn 1996; Hart et al. 1994).

Lankes et al. (2007) expanded the application of λ as a gene delivery vector by executing phage-mediated gene transfer in vivo. This group constructed recombinant λ particles encoding firefly luciferase (*luc*) in order to visualize gene delivery in real-time via the use of bioluminescence imaging (Lankes et al. 2007). They fused a $\alpha_v\beta_3$ (CD51/CD61) receptor integrin-binding peptide to gpD to increase its uptake by receptor-mediated endocytosis. This peptide, the tenth human fibronectin type III domain, was chosen because it is known to play a role in the binding/internalization in a number of mammalian viruses and it had been used to enhance the targeting of modified viral vectors to a variety of cells including professional antigen-presenting dendritic cells. The study showed preferential internalization of fused phage over their non-fused counterparts, where internalization decreased significantly in a dose-dependent fashion.

Integrin receptors are also overexpressed on cancer cells. To exploit this, Choi et al. (2014) modified filamentous phage M13 to controllably display the integrin binding motif RGD (Arg-Gly-Lys), a widely used cancer targeting peptide, on either the minor coat protein pIII or the major coat protein pVIII (Choi et al. 2014). The study examined the display of circular and linear RGD, resulting in the display of 5 or 2700 copies of linear RGD in comparison to 140 copies of circular RGD and found that the display of circular RGD facilitated more than threefold internaliza-tion of phage in HeLa cells when compared to the display of linear RGD. This highlights the importance of not only choosing a suitable targeting ligand, but also the necessary considerations of ligand display conformation.

Vaccaro et al. (2006) noted an increasing pattern of internalization following the addition of competitor proteins, indicative of a receptor-mediated process. The recombinant phage outperformed the control cells both in vitro and in vivo. In addition, they noted a 100-fold increase in phage internalization into integrin-positive versus control cells, but only a 3-fold increase in phage-mediated gene expression. This indicated that the level of internalization is not necessarily comparable to the successful delivery of genetic material (Vaccaro et al. 2006). Overall, this study provided a proof-of-concept for the use of recombinant phage to increase gene transfer in vivo, and a compelling argument for the use of phages in transgene delivery (Lankes et al. 2007; Vaccaro et al. 2006). However, it also underscores the importance of integrating mechanisms for overcoming intracellular barriers past cell entry in order to successfully deliver a genetic therapeutic.

Zanghi et al. (2007) further explored construct design, where they attempted to improve phage-mediated mammalian gene delivery of a luciferase gene through the simultaneous fusion of proteins to both the head and tail of phage λ (Zanghi et al. 2007). Multiple intracellular barriers such as cell attachment, cytoplasmic entry, endosomal escape, uncoating and nuclear import, must be overcome for successful gene transfer (Seow and Wood 2009). Multiple peptides could theoretically be displayed *in tandem*, where each peptide could function to circumvent a separate barrier. This research group reported, on average, the simultaneous display of two separate peptides: ~ 100 copies per phage particle to gpV of a CD40-binding peptide to facilitate endocytic uptake and ~ 400 fusions per phage particle to gpD of a ubiquitinylation motif to enhance intracellular trafficking of the internalized phage (Zanghi et al. 2007). Display of both the CD40-binding motif and the ubiquitinylation motif improved gene expression two-fold over display of either the CD40-binding motif or ubiquitinylation motif alone.

Although numerous methods have been developed for gene delivery, an efficient platform for protein delivery *in tandem* with gene delivery does not currently exist. Recently, Tao et al. (2013) developed a T4 DNA packaging machine using T4-based "progene" nanoparticles that were targeted to antigen-presenting cells and were expressed both in vitro and in vivo. The group fused DNA molecules onto the T4 major capsid proteins, Soc and Hoc, that would later be displayed on the phage heads. Foreign cell penetration peptides (CPPs) and proteins (β-galactosidase, dendritic cell specific receptor 205 monoclonal antibody, and CD40 ligand) were chosen for display onto Hoc. The encapsidated DNA included *gfp* (green fluorescent protein) and *luc* (luciferase) genes to enable quantifiable expression within mammalian cells. Overall, the group demonstrated efficient in vitro and in vivo progene delivery and expression of self-replicating genes into mammalian cells. While promising, further investigation particularly with respect to in vivo studies is warranted, as the strongest luciferase signal in this study was unexpectedly generated in mice treated with nanoparticles that did not display targeting ligands. The authors have attributed this finding to the migration of the targeted cells to other parts of the body, an inference that will require further investigation (Tao et al. 2013).

3   Phages as Cytotoxic Agents in Eukaryotes

While the application of phage as gene delivery vehicles could be employed to restore a gene defect or alter the physiology of the host cell, phage could further prove therapeutic as cytotoxic agents. Tissues such as tumour cells or unwanted white fat cells can be deleterious to the mammalian host and beneficially targeted for reduction and removal. Toward such therapies, the physicochemical attributes of phages can be exploited to direct cytotoxic outcomes. In a targeted manner, phage can be manipulated to deliver toxins to targeted cells (Vilchez and Jacoby 2004), or alternatively employed to stimulate immune responses and clearance of unwanted cells (Ahmadvand et al. 2011; Clark et al. 2012). While the former strategy is greatly dependent upon precise targeting and uptake of the phage particle to kill or inactivate the host cell, the latter relies upon the natural adjuvant properties of the phage to stimulate and confer immunogenicity against targeted cells (Gamage et al. 2009).

The evolution of phages as therapeutics in these capacities requires focused targeting, typically facilitated through phage display technologies. Fibroblast growth factor, in particular, has enabled the specific targeting of cancer cells with the appropriate receptors (Haq et al. 2012; Hart et al. 1994; Sperinde et al. 2001). In a series of studies by Yacoby, Benhar and others, cytotoxic bacteriophages were designed from a template of their previous cytotoxic phage used for the treatment against bacterial pathogens (Yacoby et al. 2006, 2007). This group developed a filamentous phage that was engineered to display a eukaryotic cell-binding ligand conjugated to a the cytotoxic drug, either hygromycin or doxorubicin, to be released within the targeted tumor cells. These engineered phages were shown to be effectively endocytosed, resulting in the preferential release of the cytotoxin in targeted cells (Bar et al. 2008). In another study by Chung et al. (2008), tumour cells derived from Hodgkin's-derived cell lines were targeted for apoptosis by antibody-targeted phage particles (Chung et al. 2008). Their proof-of-principle study employed an in vitro GFP expression system as a measure of phage uptake, based on the premise that efficient expression of GFP could be replaced with the expression of a cytotoxic agent in the future (Chung et al. 2008). In a similar study, Eriksson et al. (2007, 2009) also used filamentous phage to target tumour cells. However, these studies differed where the delivered phage were designed to target the host cells for removal by the host immune system without carrying a cytotoxic agent (Eriksson et al. 2007, 2009).

4   Phages for Delivery to the Central Nervous System

Delivery of therapeutics to the brain and the central nervous system (CNS) remains a challenging problem due to its complex structure, sensitivity to toxicity, and the impermeability of the blood-brain barrier (BBB). Gene delivery to the CNS has been achieved with some degree of success through direct injection into the eye and/or the cerebral spinal fluid, or direct implantation of transduced cells into brain parenchyma (Davidson and Breakefield 2003; Hampl et al. 2000); however, such methods are highly invasive, have limited penetration, and can be traumatic to the neural tissue. Overall, the capability to pass the blood brain barrier and penetrate heterogeneous neural tissue is highly desirable in a CNS-targeted therapeutic. Phage have been observed to exhibit this ability (Dabrowska et al. 2005; Frenkel and Solomon 2002) and may therefore be exploited for CNS drug and gene delivery.

Drug addiction is an important health and social problem world-wide, a prevailing culprit of which is the highly addictive recreational drug cocaine. It has been previously shown that protein-based therapeutics designed to bind to cocaine can reduce the drug load and attenuate its psychoactive effects. However, this strategy has not generally demonstrated significant therapeutic value due to the inability of these cocaine-binding proteins to cross the BBB and gain access to the CNS. To address this issue, Carrera et al. (2004) engineered a filamentous bacteriophage displaying cocaine sequestrating antibodies on its surface which blocks this drug in the brain. The modified phages were administered intranasally to rats twice a day for three consecutive days before the brains were examined, confirming the presence of the phage. The results of this study highlights the potential for phage to serve as a new system for treatment of cocaine addiction as well as serving as a platform for treatment of drug abuse (Clark and March 2006; Dickerson et al. 2005).

More recently, filamentous phages have also been demonstrated to accumulate in gliobastoma after intranasal delivery (Dor-On and Solomon 2015), potentiating their use in the treatment of brain tumours and other brain malignancies. Phage display has also been useful in the identification of several ligands capable of bypassing the BBB and targeting neural tissue (Li et al. 2011; Wan et al. 2009), which can functionalize other non-viral vectors.

5   Conclusions

Bacteriophages have evolved the natural ability to efficiently carry and deliver a genetic payload to their natural host cells—a function that continues to be exploited in the development of highly efficient, engineered phage delivery systems that can specifically target and modify non-natural host targets. The ability of the phage to cross the BBB makes it an attractive vector against neural malignancies. One major area of improvement for phage gene delivery lies in enhancing its ability to traverse intracellular barriers: notably, transport across the plasma membrane and escape from the endosomal pathway. Viral peptides such as the adenoviral penton base have been shown to mediate entry, attachment and endosomal release (Haq et al. 2012; Piersanti et al. 2004) and can be conjugated to the phage through phage display. Similarly, the protein transduction domain of HIV (TAT protein) and the simian virus 40 (SV40) T antigen nuclear localization signal have also been used to enhance the uptake and nuclear targeting of phages (Haq et al. 2012; Nakanishi et al. 2003). Additional future improvements to phage delivery technologies may exploit the display of DNA reducing DNase II inhibitor to protect DNA (Haq et al. 2012; Sperinde et al. 2001).

References

Abedon, S. T. (2009). Kinetics of phage-mediated biocontrol of bacteria. *Foodborne Pathogens and Disease, 6*(7), 807–815.

Abedon, S. T., Kuhl, S. J., Blasdel, B. G., & Kutter, E. M. (2011). Phage treatment of human infections. *Bacteriophage, 1*(2), 66–85.

Ahmadvand, D., Rahbarizadeh, F., & Moghimi, S. M. (2011). Biological targeting and innovative therapeutic interventions with phage-displayed peptides and structured nucleic acids (aptamers). *Current Opinion in Biotechnology, 22*(6), 832–838.

Bakhshinejad, B., Karimi, M., & Sadeghizadeh, M. (2014). Bacteriophages and medical oncology: Targeted gene therapy of cancer. *Medical Oncology (Northwood, London, England), 31*(8), 110.

Bar, H., Yacoby, I., & Benhar, I. (2008). Killing cancer cells by targeted drug-carrying phage nanomedicines. *BMC Biotechnology, 8*, 37.

Carrera, M. R. A., Kaufmann, G. F., Mee, J. M., Meijler, M. M., Koob, G. F., Janda, K. D. (2004). Treating cocaine addiction with viruses. *Proceedings of the National Academy of Sciences of the United States of America, 101*(28), 10416–10421. doi: 10.1073/pnas.0403795101

Choi, D. S., Jin, H., Yoo, S. Y., & Lee, S. (2014). Cyclic RGD peptide incorporation on phage major coat proteins for improved internalization by HeLa Cells. *Bioconjugate Chemistry, 25*(2), 216–223.

Chung, Y.-S. A., Sabel, K., Krönke, M., & Klimka, A. (2008). Gene transfer of Hodgkin cell lines via multivalent anti-CD30 scFv displaying bacteriophage. *BMC Molecular Biology, 9*(1), 37.

Clark, J. R., Abedon, S. T., & Hyman, P. (2012). Phages as therapeutic delivery vechicles. In *Bacteriophages in health and disease* (pp. 86–95). American Society for Microbiology.

Clark, J. R., Bartley, K., Jepson, C. D., Craik, V., & March, J. B. (2011). Comparison of a bacteriophage-delivered DNA vaccine and a commercially available recombinant protein vaccine against hepatitis B. *FEMS Immunology and Medical Microbiology, 61*(2), 197–204.

Clark, J. R., & March, J. B. (2006). Bacteriophages and biotechnology: Vaccines, gene therapy and antibacterials. *Trends in Biotechnology, 24*(5), 212–218.

Dabrowska, K., Switała-Jelen, K., Opolski, A., Weber-Dabrowska, B., & Gorski, A. (2005). Bacteriophage penetration in vertebrates. *Journal of Applied Microbiology, 98*(1), 7–13.

Davidson, B. L., & Breakefield, X. O. (2003). Neurological diseases: Viral vectors for gene delivery to the nervous system. *Nature Reviews Neuroscience, 4*(5), 353–364.

Dickerson, T. J., Kaufmann, G. F., & Janda, K. D. (2005). Bacteriophage-mediated protein delivery into the central nervous system and its application in immunopharmacotherapy. *Peptides, Proteins and Antisense, 5*(6), 773–781.

Dor-On, E., & Solomon, B. (2015). Targeting glioblastoma via intranasal administration of Ff bacteriophages. *Frontiers in Microbiology, 6*, 530.

Dunn, I. S. (1996). Mammalian cell binding and transfection mediated by surface-modified bacteriophage lambda. *Biochimie, 137*(1838), 37.

Eguchi, A., Akuta, T., Okuyama, H., Senda, T., Yokoi, H., Inokuchi, H., ... Nakanishi, M. (2001). Protein transduction domain of HIV-1 Tat protein promotes efficient delivery of DNA into mammalian cells. *The Journal of Biological Chemistry, 276*(28), 26204–26210.

Eriksson, F., Culp, W. D., Massey, R., Egevad, L., Garland, D., Persson, M. A. A., et al. (2007). Tumor specific phage particles promote tumor regression in a mouse melanoma model. *Cancer Immunology, Immunotherapy: CII, 56*(5), 677–687.

Eriksson, F., Tsagozis, P., Lundberg, K., Parsa, R., Mangsbo, S. M., Persson, M. A. A., ... Pisa, P. (2009). Tumor-specific bacteriophages induce tumor destruction through activation of tumor-associated macrophages. *Journal of Immunology (Baltimore, Md. : 1950), 182*(5), 3105–3111.

Frenkel, D., & Solomon, B. (2002). Filamentous phage as vector-mediated antibody delivery to the brain. *Proceedings of the National Academy of Sciences of the United States of America, 99*(8), 5675–5679.

Gamage, L. N. A., Ellis, J., & Hayes, S. (2009). Immunogenicity of bacteriophage lambda particles displaying porcine Circovirus 2 (PCV2) capsid protein epitopes. *Vaccine, 27*(47), 6595–6604.

Haq, I. U., Chaudhry, W. N., Akhtar, M. N., Andleeb, S., & Qadri, I. (2012). Bacteriophages and their implications on future biotechnology: A review. *Virology Journal, 9*(1), 9.

Hart, S. L., Knight, a M., Harbottle, R. P., Mistry, A., Hunger, H. D., Cutler, D. F., ... Coutelle, C. (1994). Cell binding and internalization by filamentous phage displaying a cyclic Arg-Gly-Asp-containing peptide. *The Journal of Biological Chemistry, 269*(17), 12468–12474.

Hampl, J. A., Brown, A. B., Rainov, N. G., & Breakefield, X. O. (2000). Methods for gene delivery to neural tissue. In H. R. Chin & S. O. Moldin (Eds.), *Methods in genomic neuroscience* (pp. 229–266). Boca Raton, Florida: CRC Press.

Jepson, C. D., & March, J. B. (2004). Bacteriophage lambda is a highly stable DNA vaccine delivery vehicle. *Vaccine, 22*(19), 2413–2419.

Kim, K.-P., Cha, J.-D., Jang, E.-H., Klumpp, J., Hagens, S., Hardt, W.-D., ... Loessner, M. J. (2008). PEGylation of bacteriophages increases blood circulation time and reduces T-helper type 1 immune response. *Microbial Biotechnology, 1*(3), 247–257.

Lam, A. P., & Dean, D. A. (2010). Progress and prospects: Nuclear import of nonviral vectors. *Gene Therapy, 17*(4), 439–447.

Lankes, H. A., Zanghi, C. N., Santos, K., Capella, C., Duke, C. M. P., & Dewhurst, S. (2007). In vivo gene delivery and expression by bacteriophage lambda vectors. *Journal of Applied Microbiology, 102*(5), 1337–1349.

Larocca, D., & Baird, A. (2001). Receptor-mediated gene transfer by phage-display vectors: Applications in functional genomics and gene therapy. *Drug Discovery Today, 6*(15), 793–801.

Larocca, D., Kassner, P. D., Witte, A., Ladner, R. C., Pierce, G. F., & Baird, A. (1999). Gene transfer to mammalian cells using genetically targeted filamentous bacteriophage. *The FASEB Journal, 13*, 727–734.

Li, J., Feng, L., Fan, L., Zha, Y., Guo, L., Zhang, Q., ... Wen, L. (2011). Targeting the brain with PEG-PLGA nanoparticles modified with phage-displayed peptides. *Biomaterials, 32*(21), 4943–450.

Merril, C. R., Biswas, B., Carltont, R., Jensen, N. C., Creed, G. J., Zullo, S., et al. (1996). Long-circulating bacteriophage as antibacterial agents. *Proceedings of the National Academy of Sciences, 93*, 3188–3192.

Miller, A. M., & Dean, D. A. (2009). Tissue-specific and transcription factor-mediated nuclear entry of DNA. *Advanced Drug Delivery Reviews, 61*(7–8), 603–613.

Molenaar, T. J. M., Michon, I., de Haas, S. A. M., van Berkel, T. J. C., Kuiper, J., & Biessen, E. A. L. (2002). Uptake and processing of modified bacteriophage M13 in mice: Implications for phage display. *Virology, 293*(1), 182–191.

Nakanishi, M., Eguchi, A., Akuta, T., Nagoshi, E., Fujita, S., Okabe, J., ... Hasegawa, M. (2003). Basic peptides as functional components of non-viral gene transfer vehicles. *Current Protein and Peptide Science, 4*(2), 141–150.

Nayak, S., & Herzog, R. W. (2010). Progress and prospects: Immune responses to viral vectors. *Gene Therapy, 17*(3), 295–304.

Nicastro, J., Sheldon, K., El-Zarkout, F. A., Sokolenko, S., Aucoin, M. G., & Slavcev, R. (2013). Construction and analysis of a genetically tuneable lytic phage display system. *Applied Microbiology and Biotechnology, 97*(17), 7791–7804.

Nicastro, J., Sheldon, K., & Slavcev, R. A. (2014). Bacteriophage lambda display systems: Developments and applications. *Applied Microbiology and Biotechnology, 98*(7), 2853–2866.

Piersanti, S., Cherubini, G., Martina, Y., Salone, B., Avitabile, D., Grosso, F., ... Saggio, I. (2004). Mammalian cell transduction and internalization properties of lambda phages displaying the full-length adenoviral penton base or its central domain. *Journal of Molecular Medicine (Berlin, Germany), 82*(7), 467–476. http://doi.org/10.1007/s00109-004-0543-2

Poul, M. A., & Marks, J. D. (1999). Targeted gene delivery to mammalian cells by filamentous bacteriophage. *Journal of Molecular Biology, 288*(2), 203–211.

Prieto, Y., & Sánchez, O. (2007). Self-complementary sequences induce the formation of double-stranded filamentous phages. *Biochimica et Biophysica Acta—General Subjects, 1770*(8), 1081–1084.

Seow, Y., & Wood, M. J. (2009). Biological gene delivery vehicles: Beyond viral vectors. *Molecular Therapy: The Journal of the American Society of Gene Therapy, 17*(5), 767–777.

Smith, G. P., & Petrenko, V. A. (1997). Phage display. *Chemical Reviews, 2665*(96), 391–410.

Somia, N., & Verma, I. M. (2000). Gene therapy: Trials and tribulations. *Nature Reviews Genetics, 1*(2), 91–99.

Specthrie, L., Bullitt, E., Horiuchi, K., Model, P., Russel, M., & Makowski, L. (1992). Construction of a microphage variant of filamentous bacteriophage. *Journal of Molecular Biology, 228*(3), 720–724. http://doi.org/10.1016/0022-2836(92)90858-H

Sperinde, J. J., Choi, S. J., & Szoka, F. C. (2001). Phage display selection of a peptide DNase II inhibitor that enhances gene delivery. *The Journal of Gene Medicine, 3*(2), 101–108.

Tao, P., Mahalingam, M., Marasa, B. S., Zhang, Z., Chopra, A. K., & Rao, V. B. (2013). In vitro and in vivo delivery of genes and proteins using the bacteriophage T4 DNA packaging machine. *Proceedings of the National Academy of Sciences of the United States of America, 110*(13), 4–9. h.

Trabulo, S., Cardoso, A. L., Cardoso, A. M. S., Düzgüneş, N., Jurado, A. S., & de Lima, M. C. P. (2012). Cell-penetrating peptide-based systems for nucleic acid delivery: A biological and biophysical approach. *Methods in Enzymology, 509*, 277–300.

Vaccaro, P., Pavoni, E., Monteriù, G., Andrea, P., Felici, F., & Minenkova, O. (2006). Efficient display of scFv antibodies on bacteriophage lambda. *Journal of Immunological Methods, 310*(1–2), 149–158.

Vilchez, S., & Jacoby, J. (2004). Display of biologically functional insecticidal toxin on the surface of lambda phage. *Applied and Environmental, 70*(11), 6587–6594.

Wan, X.-M., Chen, Y.-P., Xu, W.-R., Yang, W., & Wen, L.-P. (2009). Identification of nose-to-brain homing peptide through phage display. *Peptides, 30*(2), 343–350.

Willats, W. G. T. (2002). Phage display: Practicalities and prospects. *Plant Molecular Biology, 50*, 837–854.

Yacoby, I., Bar, H., & Benhar, I. (2007). Targeted drug-carrying bacteriophages as antibacterial nanomedicines. *Antimicrobial Agents and Chemotherapy, 51*(6), 2156–2163.

Yacoby, I., & Benhar, I. (2008). Targeted filamentous bacteriophages as therapeutic agents. *Expert opinion on drug delivery, 5*(September), 321–329.

Yacoby, I., Shamis, M., Bar, H., Shabat, D., & Benhar, I. (2006). Targeting antibacterial agents by using drug-carrying filamentous bacteriophages. *Antimicrobial Agents and Chemotherapy, 50*(6), 2087–2097.

Yokoyamakobayashi, M., & Kato, S. (1993). Recombinant f1 phage particles can transfect monkey COS-7 Cell by DEAE dextran method. *Biochemical and Biophysical Research Communications, 192*(2), 935–939.

Yokoyamakobayashi, M., & Kato, S. (1994). Recombinant f1 phage-mediated transfection of mammalian cells using lipopolyamine technique. *Analytical Biochemistry, 223*(1), 130–134.

Zanghi, C. N., Sapinoro, R., Bradel-Tretheway, B., & Dewhurst, S. (2007). A tractable method for simultaneous modifications to the head and tail of bacteriophage lambda and its application to enhancing phage-mediated gene delivery. *Nucleic Acids Research, 35*(8), e59.

Chapter 5
Phage Probiotics

Abstract The human body is composed of more than just human cells. The microbiota encompasses a large diversity of bacteria, eukarya, archaea, and viruses that reside in human organs. It has received increasingly more attention as its roles in human homeostasis and development have been identified and as such, probiotics are growing in popularity as a means to improve health. The gut microbiota protects against pathogens, contributes to energy metabolism, mediates the immune system, and modulates tissue development. As bacteriophage are natural predators of bacteria, they may be well suited to manipulate the gut microbiota. Phage predation naturally impacts gut population dynamics and influences microbial gene expression. Here, we review some key aspects of gut microbial and phage function, and explore new advances in phage biotechnology that may pave the way for future phage probiotic applications.

1 Introduction: The Gut Microbiota and Probiotics

The number of microorganisms inhabiting the human body greatly exceeds that of host cells (Turnbaugh et al. 2007). These residents form highly complex and mutualistic microbial ecological niches within the human host and thus, interest in their roles in the management of their hosts' health has increased in recent years (Blaser et al. 2013). Bacteria and their viral predators, bacteriophages (phages), are the most prevalent inhabitants of the mammalian gastrointestinal system (Marchesi and Shanahan 2007), the densest and most complex ecosystem in the body. The gut microbiota or gut "flora" comprises all entities including bacteria, archaea, and eukarya, and viruses residing within the gastrointestinal tract (De Paepe et al. 2014). These residents are often referred to as "commensal microorganisms" or simply, the "commensals". Diverse species colonize the colon, though their impact, both positive and negative, on their host has only recently begun to be investigated. This chapter is a preliminary review of the commensal and probiotic bacteria and proposes how bacteriophages can accomplish or supplement similar health outcomes.

© The Author(s) 2016
J. Nicastro et al., *Bacteriophage Applications—Historical Perspective and Future Potential*, SpringerBriefs in Biochemistry and Molecular Biology,
DOI 10.1007/978-3-319-45791-8_5

The gut is home to an extraordinarily large microbial community: microbial density here exceeds 10^{12} cells per gram of feces (Macpherson and Harris 2004). While highly diverse (Qin et al. 2010), *Bacteriodetes* and *Firmicutes* tend to dominate this domain in humans (Huttenhower et al. 2012; Qin et al. 2010); anaerobic bacteria far outnumber their aerobic counterparts. An individual's gut flora begins flourishing at birth (Martin et al. 2010; Wassenaar and Panigrahi 2014) and fluctuates in composition as the host ages, but remains unique to each individual (Sommer and Bäckhed 2013; Turnbaugh et al. 2009). Overall gut microbial composition varies geographically (Yatsunenko et al. 2012) but also strongly depends on both genetic and environmental factors (Huttenhower et al. 2012; Turnbaugh et al. 2009; Yatsunenko et al. 2012). Research into the gut microbiota has greatly expanded in the last decade (Turnbaugh et al. 2007), but we still lack a deep understanding of how the microbiome functions.

The microbiota is sensitive to changes in host behaviour, including diet (Ding et al. 2010; Turnbaugh et al. 2009), antibiotic use (Modi et al. 2014), and lifestyle (Martínez et al. 2015), so alterations to any of these factors can result in rapid fluctuations. "Dysbiosis", or perturbations in "normal" gut microbial composition, is associated with some intestinal diseases (Stecher et al. 2013), in particular: inflammatory bowel disease (IBD) (Tamboli et al. 2004; Young et al. 2011), irritable bowel syndrome (IBS) (Shanahan 2012), Crohn disease, and ulcerative colitis (Shanahan 2012). Diet-related imbalances in the gut microbiota are also associated with obesity (Cani 2013; Delzenne and Cani 2011; Ding et al. 2010; Joyce et al. 2014), where obesity-related disorders can lead to type II diabetes (Cani et al. 2012; Cani 2008; Qin et al. 2012). The question arises whether or not the connection between the microbiome and disease is causative or merely correlative (Cho and Blaser 2012). There is a strong rationale for the manipulation of gut flora as a therapeutic measure.

Defined by the World Health Organization as "live microorganisms which when administered in adequate amounts confer a health benefit on the host" (Hill et al. 2014), probiotics modify the landscape of the gut microbiota in order to influence human health. However, while probiotics are now widely available and establishing a niche in the food market, the lack of regulation and explosive growth despite poor supporting evidence has contributed to extensive misinformation (Senok et al. 2005). It should be stressed that the use of probiotics be approached cautiously.

To date, probiotics have been positively linked to the treatment of gastrointestinal diseases and disorders including, perhaps most importantly, antibiotic-associated diarrhea (AAD) (Hempel et al. 2012; Hungin et al. 2013; Szajewska et al. 2006; Videlock and Cremonini 2012). Their use also appears to be correlated to the alleviation of other forms of acute diarrhea (Applegate et al. 2013; McFarland 2007; Sazawal et al. 2006), infections by *Clostridium difficile* (Johnston et al. 2012; Ritchie and Romanuk 2012; Videlock and Cremonini 2012) or *Helicobacter pylori* (Dang et al. 2014; Hungin et al. 2013; Ritchie and Romanuk 2012), IBS (Hoveyda et al. 2009; Hungin et al. 2013), and constipation (Dimidi et al. 2014). Use of probiotics also positively correlates with the prevention and/or treatment of other diseases, particularly during child development such as: necrotizing enterocolitis (AlFaleh and Anabrees 2014), hepatic encephalopathy (Holte et al. 2012), upper respiratory tract infections (Hao et al. 2011), atopic dermatitis (Kalliomäki et al. 2003; Pelucchi et al. 2012), atopic eczema (Doege et al. 2012), and ventilator-associated pneumonia (Petrof et al. 2012). Overall, there is evidence supporting the use of probiotics for certain diseases such as AAD, IBD, and some gut infections. While there exists some support for overall improvement of health with probiotics (Hungin et al. 2013), the lack of regulation (JAMA 2014) of the probiotic market and availability of well-controlled evidence makes it difficult to judge their true effect in healthy individuals. The elucidation of the mechanisms behind their proposed benefits is vital to using them to their fullest extent.

2 Roles of the Gut Microbiota and Probiotics

2.1 Protection Against Pathogens

The intestinal epithelium physically separates the contents of the intestinal lumen from the rest of the body. It is protected by a mucus layer secreted by goblet cells, approximately 100–200 μm thick, and composed of mucin glycoproteins (Pullan et al. 1994). The epithelial cells also secrete antimicrobial peptides such as defensins, which limit bacterial penetration into the epithelial layer. However, this physical surface is also ripe for bacterial adherence and consequently, the formation of biofilms and local micro-ecosystems. Nonpathogenic microorganisms in the gut adhere to the gut mucus and thereby reduce pathogen access to host epithelial cells.

The density of the gut microbiota is also thought to provide protection against colonization by opportunistic pathogens since the beneficial bacteria would be able to outcompete potential pathogens for space and metabolic resources, commonly referred to as "colonization resistance" (Abt and Pamer 2014). Some microorganisms produce mucin (Barr et al. 2013a; Mack et al. 1999), which also inhibits other bacteria from adhering to surfaces. Some probiotic bacteria also exert direct antimicrobial effects through the release of bacteriocins, which are highly specific antimicrobial molecules (Corr et al. 2007).

2.2 Metabolism

Energy metabolism is closely linked to each individual's microbiome. Mammals, particularly ruminants, require the gut microbiota to process dietary polysaccharides like cellulose and dietary fibre (Bergman 1990; Flint et al. 2008; Wolin 1981). Microbial fermentation in the gut (Hamer et al. 2008; Wolin 1981) produces energy resources for epithelial cells here. The gut flora contributes to acetogenesis (Rey et al. 2010) and methanogenesis (Vanderhaeghen et al. 2015) metabolic processes. These microorganisms have also been found to metabolize drugs or drug metabolic byproducts (Goldman et al. 1974), which adds another important consideration in the design of pharmaceuticals. Metabolites produced by the gut microbiota may be both harmful and beneficial to the host (Louis et al. 2014).

The role of the gut flora in obesity has also been well established (Cani 2013; Delzenne and Cani 2011; Turnbaugh et al. 2009). Microbiome composition has been shown to differ between obese and lean twins, although the high variability between individuals makes it difficult to identify key species related to obesity (Turnbaugh et al. 2009). Increased energy storage has been attributed to the suppression of fasting-induced adipose factor (Fiaf) expression by the gut flora. Bäckhed et al. (2004, 2007) suggested that elevated Fiaf expression and increased AMP-activated protein kinase (AMPK) expression in skeletal muscle and liver tissue in the absence of the gut flora could also contribute to increased fatty acid metabolism.

2.3 Immunomodulation

Humans benefit from the ability of the microbiota to protect against pathogens and to metabolize otherwise indigestible products, while the microbiota benefits from the nutrient-rich ecosystem provided by the host. Mammals have therefore adapted to their microbial passengers (Lee and Mazmanian 2010) as they are able to symbiotically co-exist without mounting inflammatory responses against commensals and still rapidly mount immune responses against invading pathogens. Immune tolerance in the gut appears to be highly dependent on host regulatory T cells (Izcue et al. 2006) and epithelial Toll-like Receptor (TLR) signaling (Hornef and Bogdan 2005). Recent investigations into the influence of the gut microbiota on the host immune system have revealed important roles in immune cell development and differentiation. The expansive role of the gut flora in immune development has been thoroughly reviewed (Hill and Artis 2010; Kranich et al. 2011).

It is clear from studies in germ-free mice that the mucosal immune system is underdeveloped in the absence of the gut flora (Macpherson and Harris 2004). Germ-free mice have smaller Peyer's patches, smaller lymph nodes, fewer lamina propria, and diminished numbers of IgA producing cells (Macpherson et al. 2001). Both the gut and probiotic bacteria maintain gut health through modulation of the host immune system at the gut mucosal surface and also through secretion of immunomodulatory molecules (Atarashi et al. 2008; Mazmanian et al. 2008). Past the epithelial barrier, gut bacteria are taken up by intestinal macrophages in the lamina propria much like their pathogenic counterparts, but they can persist in intestinal dendritic cells (DCs) and induce local protective IgA production (Macpherson and Uhr 2004). Probiotics can inhibit the immune cell apoptotic pathway (Yan and Polk 2002; Yan et al. 2007), induce cytokine production

(Grangette et al. 2005; Smits et al. 2005), and also enhance the overall innate immune response in the gut, thereby increasing immune cell longevity and pathogen colonization resistance (Grangette et al. 2005; Wehkamp et al. 2004).

2.4 Tissue Development and Maintenance

Alongside protection against pathogens, probiotic and gut bacteria may also have a role in maintaining the gut mucosal surface. The gut flora modulates gene expression in the intestinal epithelium and therefore may influence cell differentiation and development. Mucin production by probiotic and resident gut bacteria contributes to the structure of the gut mucus layer. The gut microbiota may also stimulate angiogenesis in the intestinal villi (Stappenbeck et al. 2002), promoting development of microvasculature. Murine studies have further shown a role for the gut microbiota in the regulation of bone mass (Sjögren et al. 2012). The authors suggest that the gut microbiota stimulates cytokine expression in the bone, which modulates osteoclast-mediated bone resorption thereby influencing bone mass.

Studies in germ-free models have greatly aided in dissecting the microbiota's influence over the human body. Additionally, microarray studies have also highlighted the wide range of gene expression changes associated with the colonization of the gut flora (Hooper et al. 2001). The microbial "organ" clearly plays an astonishingly vital role in human homeostasis.

3 Bacteriophages in the Gut

Bacteriophages are the most abundant and diverse biological entities on the planet, numbering at least ten phages per microbial cell in almost every environment, including the gut. Thousands of virotypes exist per fecal sample and many still remain uncharacterized (Breitbart et al. 2003; Kim et al. 2011). Even now, new species are still being discovered in the gut by advancing metagenomic strategies (Dutilh et al. 2014). Viral genomes are highly abundant and well represented in the gut microbiome (Abeles and Pride 2014; Breitbart et al. 2003; De Paepe et al. 2014; Minot et al. 2011; Modi et al. 2013; Reyes et al. 2010), where the majority of these are members of the *Siphoviridae* family (Breitbart et al. 2003), followed by *Podoviridae* (Kim et al. 2011; Minot et al. 2011; Reyes et al. 2010). Gut viral particles are present in lower abundance in comparison to marine environments, which suggests that the gut "virome" is likely dominated by temperate rather than lytic phages (Reyes et al. 2010, 2012). The virome, like its microbial counterpart, is unique to each individual (Reyes et al. 2010); it begins development at birth (Breitbart et al. 2008), and fluctuates in response to host behaviour (Minot et al. 2011; Reyes et al. 2010).

3.1 Phage Population Dynamics

The diversity of the gut virome ensures coverage of a very broad host range. Nonetheless, the majority of free gut phages will most likely target the gut microbiota. In any environment, phages are highly effective at infecting their prey. However, phage predation has yet to drive bacterial species to extinction; both bacteria and viruses are highly proliferative in the gut environment, for example. The study of phage-prey population dynamics has generated many different models to explain the co-existence of phage and their bacterial prey. In all these models, phage predation is shown to be a driving force behind bacterial diversity.

The co-evolutionary arms race model, where bacteria escape phage infection by evolving resistance mechanisms that are subsequently circumvented by phage evolution, is termed "Red Queen dynamics". It is often used to describe how phage mediate bacterial diversity, with the CRISPR/Cas system serving as the clearest example (Makarova et al. 2011). However, Red Queen dynamics alone cannot explain the persistence and stability of co-existing phage and bacterial populations (Heilmann et al. 2012). A study in *Vibrio cholera* observed that mutations for phage resistance negatively impacted those mutants' competitive fitness against their phage-sensitive counterparts in the same environment (Seed et al. 2014). Multiple selective pressures are at play.

Phage-microbial ecosystems are generally thought to exhibit "kill-the-winner" dynamics, where phage population growth closely follows microbial expansion as phages prey on the largest host populations (Rodriguez-Valera et al. 2009; Thingstad 2000; Wiggins and Alexander 1985). As the corresponding host population dwindles from phage infection, new "winners" increase in population, thereby ensuring microbial diversity in the environment. Similarly, the "constant diversity" (CD) model strives to explain the maintenance of diversity even among closely related lineages (Rodriguez-Valera et al. 2009), and has been demonstrated in coliphages. Phages infect bacterial cells by targeting cell surface receptors. According to CD dynamics, the presence of any new receptor, which is analogous to a new mutant cell entering the ecosystem, would be followed by a corresponding increase in the targeting phage population. In essence, these dynamics illustrate how phage predation maintains diversity within a bacterial population by ensuring rapid bacterial turnover. However, such dynamics are not expected to be overtly prevalent within the human gut, where physical constraints and extremely high bacterial density will dictate the structure of the microbial community. Spatial compartmentalization modulates phage-microbe interactions (Brüssow 2013; Macfarlane and Dillon 2007). Kill-the-winner and CD dynamics may still be relevant within the gut lumen or within micro-ecosystems along the mucus layer, but do not appear to be dominant. The "spatial refuge hypothesis" (Heilmann et al. 2012) which models phage-microbial interactions in the context of biofilms may be more pertinent. In this model, high-density bacterial populations form bacterial refuges, which are not easily penetrated by phage. Instead, phage infection is highest at the boundaries of such refuges. Both phage and prey populations stably co-exist in the long-term and the model also permits co-evolution.

Because many bacteria are actually physically inaccessible to phages, the phage-prey dynamic in the gut is thought to be influenced by prophages more so than by lytic phages (Mills et al. 2013; Reyes et al. 2012). While prophages are known for encoding virulence factors and enabling pathogenic bacteria (Keen 2015), they may impart other nontoxic competitive advantages. Prophage genes involved in anaerobic metabolism are abundant in fecal viromes (Reyes et al. 2010). One strain of *Enterococcus faecalis* maintains a niche within the gut ecosystem by producing a composite phage from two separate prophage elements (Duerkop et al. 2012). Phage production is triggered by limited nutrient availability and the composite phage attacks competing *E. faecalis*, thereby conferring a competitive benefit to the strain.

Prophage induction, the genetic shift from lysogenic cycle into the lytic, results in cell lysis and release of phage particles and can also play a role in ensuring microbial diversity. Prophage induction typically occurs under stressful conditions threatening host cell survival, presumably so the temperate phage can move on to better hosts. Within the human ecosystem, bacteria are capable of exploiting phage induction as a means to kill off competitors (Selva et al. 2009). Phage infection and prophage induction is also linked to quorum sensing (QS). QS systems regulate bacterial gene expression in response to changes in cell density (Miller and Bassler 2001) and have been linked to a number of important pathogenic phenotypes including biofilm formation (Jayaraman and Wood 2008) and toxin production (Smith and Harris 2002). QS-mediated gene expression is notable in the dense gut ecosystem. QS signaling molecules have been shown to induce prophages in soil lysogens (Ghosh et al. 2009), providing further evidence that phage production is linked with high cell density. Additionally, QS homologs have been recently discovered in the temperate phage phiCDHM1, proposed to have originated from its host, *C. difficile* but are genetically distinct (Hargreaves et al. 2014). The authors propose that phiCDHM1 may therefore evoke QS responses in its host in a novel mechanism for survival.

Phage predation and prophage induction may be problematic for probiotic administration as the phage-sensitive probiotic population runs the risk of being eliminated by gut phage upon entry into the gut ecosystem (Ventura et al. 2011). The use of phage-resistant probiotics would most likely disseminate phage resistance into the environment, again potentially upsetting the equilibrium. The gut virome is thus an important parameter to consider in designing new probiotic strategies.

3.2 Protection Against Pathogens and Immunomodulation

As the predators of bacteria, phage are capable of effectively and specifically eliminating invading pathogenic bacteria. Naturally, they can protect the gut against pathogen colonization, but the question remains as to how they can protect against all pathogens. Recently, phage that may specifically play this role were described by Barr et al. (2013a) in a model "Bacteriophage Adhering to Mucus" (BAM) (Barr et al. 2013a). Phage capsids were previously shown to have Ig-like domains, but only recently were these capsids discovered to interact with the mucin glycoproteins of the mucus layer, effectively embedding phage heads into the gut mucus layer. The tails of these embedded phages are exposed to the gut lumen and can infect susceptible bacteria near the mucosal wall. Barr et al successfully demonstrated phage embedded T4 phage activity in vitro against invading *Escherichia coli* with supporting in vivo evidence (Barr et al. 2013a). This identifies a novel function for bacteriophage in innate gut immunity (Barr et al. 2013b), though many questions still remain regarding the mechanisms of BAM-mediated killing in vivo. Furthermore, phage have been shown to activate tumor-associated macrophages via TLR signaling (Eriksson et al. 2009) and may have other roles in adaptive immune activation.

Phages are very prevalent in the gut microbiome, with demonstrated roles in controlling bacterial populations and mediating diversity within the environment. Metagenomic studies have also revealed the large scope of prophage genomes, although the mechanisms behind prophage influence over the bacterial community has yet to be fully elucidated. While the gut microbiome has been shown to play roles in human regulatory and metabolic pathways, the gut virome has been thought to mediate and control the microbiota in turn. In essence, the phages modulate the bacteria, which in turn modulate human health and homeostasis.

4 Applications of Phage

Most applications of phage biotechnology seek to apply or enhance phage-mediated killing of bacterial targets. The emergence of multi-drug resistant pathogens, as a consequence of the overwhelming application of antibiotics over the past several decades, is growing and becoming increasingly more difficult to treat. Bacteriophage antibacterial technologies have therefore begun to garner great interest as alternatives to antibiotics. We will consider some examples of phage biotechnology particularly applicable to the gut microbiota and examine how they can influence human health.

4.1 Lytic Phage Therapy

Many phage applications exploit the specificity of phage to pathogenic bacteria. Phage therapy is the use of lytic phage administered directly to a patient either locally or systemically to infect and kill targeted pathogen (Abedon et al. 2011; Thiel 2004). While originally discovered in early in the 1900s, the use of phage did not gain traction in Western medicine and was displaced by antibiotics after World War II [reviewed in (Gill and Young 2011; Sulakvelidze 2011)]. There is now reemerging interest in phage therapy as we enter the post-antibiotic era (Thiel 2004).

Phage preparations have been effectively used in food preparation to prevent contamination (Tsonos et al. 2014). In an effort to control foodborne diseases, phage therapy has long been tested as a means to prevent or treat diseases in livestock (Atterbury 2009; Greer 2005) such as cattle (Smith et al. 1987), poultry (Fiorentin 2005; Miller et al. 2010; Oliveira et al. 2009), and pigs (Jamalludeen et al. 2009). More recently, phages are also being offered as probiotics in humans, particularly as a supplement or alternative to antibiotics. There is a growing number of biopharmaceutical companies focusing on phage-based therapeutics. IntraLytix (www.intralytix.com) has developed multiple phage preparations for the direct elimination of pathogenic bacteria, several veterinary phage therapies, and a new phage therapeutic against *Shigella* intended for consumption. The George Eliava Institute of Bacteriophages, Microbiology and Virology (www.eliava-institute.org) developed "*Intestiphage*", another potential "phage probiotic", among other phage therapies.

Phage-derived antimicrobial molecules are also increasingly gaining interest. Endolysins, hydrolase enzymes that degrade the bacterial cell wall to promote lysis (Nakonieczna et al. 2015), and holins, small membrane proteins that induce lethal lesions in the cell membrane (Young 2002) are being investigated. These highly specific and small molecules can serve as excellent alternatives to common antibiotics and against drug-resistant bacteria. They may be used to circumvent resistance and antibiotic-associated disease states resulting from broad-spectrum antibiotics. Endolysins against *C. difficile* have already been identified and have been shown to be specifically active against the pathogen (Rea et al. 2013).

4.2 Phage Biotechnology

Phage display, the covalent linkage of proteins to phage capsid proteins, exploits phage specificity and may be used to selectively deliver proteins to their targets. Filamentous bacteriophage have been demonstrated to deliver such covalently linked active antibiotic moieties specifically to pathogenic *E. coli* strains (Westwater et al. 2003; Yacoby et al. 2006, 2007; Yacoby and Benhar 2008). These antibiotic molecules have the advantage of being specifically targeted to pathogens and can therefore avoid killing other potentially beneficial commensal microbiota; however, this is highly dependent on the intrinsic tropism of the phage in question. It may be possible to expand phage tropism by display of targeting peptides. This can circumvent phage resistance or target the phage to a new host.

While phage display is a very powerful technology, the delivery of the displayed moieties is limited to the number of phage particles available. Lu and Collins (2007) demonstrated an alternative two-pronged approach to dispersing biofilms, where they enhanced the natural ability of T7 phage to penetrate biofilms by encoding the gene for DspB, a EPS-degrading enzyme into the phage genome (Lu and Collins 2007). After infection, replication, and lysis by T7, DspB would be released, further degrading the biofilm matrix and exposing cells within the biofilm. The lytic phage T7 was shown to hijack target cell machinery in order to express and release recombinant proteins, further underscoring the potential of manipulating phage beyond basic phage therapy. While Lu et al. concentrated on biofilm degradation, other beneficial proteins could potentially be encoded and released as well. It may be possible to mimic the activities of probiotic bacteria through phage-mediated delivery, production, and release of relevant proteins, such as mucin or bacteriocins.

However, the release of internal bacterial cell components and large numbers of phage progeny upon lysis can have inflammatory effects for the host (Lepper et al. 2002; Merril et al. 1996). Nonlytic phage therapies may be more advantageous for some applications in the gut by avoiding undesirable secondary immune responses. These proof-of-principle experiments have generally concentrated on exploiting lysogenic and filamentous phage to deliver inducible cytotoxic genetic cargo.

Paul et al. (2011) constructed a lysis-deficient phage as a *Staphylococcus aureus* phage therapeutic (Paul et al. 2011). The temperate phage was shown to lysogenize the target pathogen in vivo by way of holin-induced lesions in the cell membrane, which depolarizes the plasma membrane and leads to cell death but not cell lysis.

The lethal agent delivery system first described by Norris et al. (2000) used non-replicative nonlytic phageheads that carried plasmids encoding antimicrobial agents (Norris et al. 2000; Westwater et al. 2003). These phageheads are generated from a defective lysogen incapable of packaging the phage genome, but capable of replicating and packaging plasmids with a bacteriophage packaging signal (phagemids). Upon delivery, these phageheads specifically adhere to the cell surface receptors of the targeted pathogen and release the lethal phagemids into the targeted cell. Expression of the antimicrobial genes results in swift cell death of the pathogen, again, independent of phage lysis. The delivery of type II endonucleases or modified phage holin genes by modified M13 phage was demonstrated in *E. coli* (Hagens and Bläsi 2003). A similar approach to deliver endonucleases was used in *Pseudomonas aeruginosa* (Hagens et al. 2004).

More recently, a different modular phagemid system for the expression of nonlytic antimicrobial peptides has been developed by Krom et al. (2015) and tested in mice (Krom et al. 2015). The phagemid carries the M13 packaging signal, which is then packaged by phage proteins expressed from a helper plasmid in the specialized production strain. Again, the resultant viral particles are devoid of phage genome and only carry the lethal phagemids, which can then be delivered to target bacteria.

Phages can modify target behaviour as well. Lu and Collins (2009) generated phage that targeted the SOS DNA repair system as a strategy to enhance antibiotic sensitivity in *E. coli* (Lu and Collins 2009). They successfully demonstrated their engineered M13 mp18 phage as an effective antibiotic adjuvant. In another study, Edgar et al. (2012) infected pathogenic antibiotic-resistant bacteria with temperate phage to reverse their resistance (Edgar et al. 2012). Streptomycin and nalidixic acid-resistant *E. coli* strains were lysogenized by λ phage carrying the alleles *rpsL* and *gyrA* in order to confer antibiotic sensitivity by replacing antibiotic targets, ribosomal and gyrase enzyme functions, respectively. Another approach to combat antibiotic resistance is phage delivery of small regulatory RNA (sRNA) (Libis et al. 2014). Libis et al. (2014) designed a phagemid encoding sRNA against dominant antibiotic resistance genes and delivered them using M13 (Libis et al. 2014). They observed re-sensitization to antibiotics kanamycin and chloramphenicol in *E. coli* after infection with the recombinant M13.

In these nonlytic prokaryotic gene delivery approaches, the use of phagemids circumvents development of phage resistance since the delivered phagemids are devoid of phage elements and can be easily modified to accommodate different genetic cargo.

5 Conclusions

Most applications of phage rely on or enhance their ability to kill pathogenic bacteria. Phage therapy is poised as a favourable alternative to antibiotics. However, unmodified phage therapy is highly susceptible to phage resistance and relies on the natural tropism of the phage. Phage biotechnologies such as phage display and both, lytic and nonlytic phage gene delivery, are areas of rich development. They can be creatively applied to not only enhance phage-mediated antibacterials by expanding target range and/or improving cytotoxicity, but also enhance commensal microbiota function. Gene delivery of cytotoxic genes may be particularly powerful, especially in the post-antibiotic era. Even so, nonlytic gene delivery also has the dangers of lateral gene transfer, phage mutation, and uncontrolled prophage induction, which must be taken into consideration.

Greater insight into the microbiome is required in general. Many health indications been linked to the gut microbiome and they may be very well suited for probiotic—bacterial or phage—treatment. Thus far, phage and bacterial therapies have proven successful as antimicrobials, particularly in animals. Further investigation into the mutualistic roles of the gut microbiome and virome will facilitate better probiotics for use in healthy individuals. Phage probiotics outside of lytic phage therapy may be very difficult to implement in humans due to lack of regulation and oversight.

References

Abedon, S. T., Kuhl, S. J., Blasdel, B. G., & Kutter, E. M. (2011). Phage treatment of human infections. *Bacteriophage*, *1*(2), 66–85. http://doi.org/10.4161/bact.1.2.15845

Abeles, S. R., & Pride, D. T. (2014). Molecular bases and role of viruses in the human microbiome. *Journal of Molecular Biology*, *426*(23), 3892–3906. http://doi.org/10.1016/j.jmb.2014.07.002

Abt, M. C., & Pamer, E. G. (2014). Commensal bacteria mediated defenses against pathogens. *Current Opinion in Immunology, 29*, 16–22. http://doi.org/10.1016/j.coi.2014.03.003

AlFaleh, K., & Anabrees, J. (2014). Probiotics for prevention of necrotizing enterocolitis in preterm infants. *Cochrane Database of Systematic Reviews, 9*(3), 584–671. http://doi.org/10. 1002/ebch.1976

Applegate, J. A., Fischer Walker, C. L., Ambikapathi, R., & Black, R. E. (2013). Systematic review of probiotics for the treatment of community-acquired acute diarrhea in children. *BMC Public Health, 13*(Suppl 3), S16. http://doi.org/10.1186/1471-2458-13-S3-S16

Atarashi, K., Nishimura, J., Shima, T., Umesaki, Y., Yamamoto, M., Onoue, M., Takeda, K., et al. (2008). ATP drives lamina propria T(H)17 cell differentiation. *Nature, 455*(7214), 808–812. http://doi.org/10.1038/nature07240

Atterbury, R. J. (2009). Bacteriophage biocontrol in animals and meat products. *Microbial Biotechnology, 2*(6), 601–612. http://doi.org/10.1111/j.1751-7915.2009.00089.x

Bäckhed, F., Ding, H., Wang, T., Hooper, L. V, Koh, G. Y., Nagy, A., Gordon, J. I., et al. (2004). The gut microbiota as an environmental factor that regulates fat storage. *Proceedings of the National Academy of Sciences, 101*(44), 15718–15723. http://doi.org/10.1073/pnas. 0407076101

Bäckhed, F., Manchester, J. K., Semenkovich, C. F., & Gordon, J. I. (2007). Mechanisms underlying the resistance to diet-induced obesity in germ-free mice. *Proceedings of the National Academy of Sciences, 104*(3), 979–984. http://doi.org/10.1073/pnas.0605374104

Barr, J. J., Auro, R., Furlan, M., Whiteson, K. L., Erb, M. L., Pogliano, J., Rohwer, F., et al. (2013a). Bacteriophage adhering to mucus provide a non–host-derived immunity. *Proceedings of the National Academy of Sciences, 110*(26), 10771–10776. http://doi.org/10.1073/pnas. 1305923110

Barr, J. J., Youle, M., & Rohwer, F. (2013b). Innate and acquired bacteriophage-mediated immunity. *Bacteriophage, 3*(3), e25857. http://doi.org/10.4161/bact.25857

Bergman, E. N. (1990). Energy contributions of volatile fatty acids from the gastrointestinal tract in various species. *Physiological Reviews, 70*(2), 567–590. Retrieved from http://physrev. physiology.org/content/70/2/567.long

Blaser, M. J., Bork, P., Fraser, C., Knight, R., & Wang, J. (2013). The microbiome explored: Recent insights and future challenges. *Nature Reviews. Microbiology, 11*(3), 213–217. http:// doi.org/10.1038/nrmicro2973

Breitbart, M., Haynes, M., Kelley, S., Angly, F. E., Edwards, R. A., Felts, B., Rohwer, F., et al. (2008). Viral diversity and dynamics in an infant gut. *Research in Microbiology, 159*(5), 367–373. http://doi.org/10.1016/j.resmic.2008.04.006

Breitbart, M., Hewson, I., Felts, B., Mahaffy, J. M., Nulton, J., Salamon, P., Rohwer, F., et al. (2003). Metagenomic analyses of an uncultured viral community from human feces. *Journal of Bacteriology, 185*(20), 6220–6223. http://doi.org/10.1128/JB.185.20.6220-6223.2003

Brüssow, H. (2013). Bacteriophage-host interaction: From splendid isolation into a messy reality. *Current Opinion in Microbiology, 16*(4), 500–506. http://doi.org/10.1016/j.mib.2013.04.007

Cani, P. D. (2008). Role of gut microflora in the development of obesity and insulin resistance following high-fat diet feeding. *Pathologie Biologie, 56*(5), 305–309. Retrieved from http:// resolver.scholarsportal.info/resolve/03698114/v56i0005/305_rogmitirfhdf.xml

Cani, P. D. (2013). Gut microbiota and obesity: Lessons from the microbiome. *Briefings in Functional Genomics, 12*(4), 381–387. http://doi.org/10.1093/bfgp/elt014

Cani, P. D., Osto, M., Geurts, L., & Everard, A. (2012). Involvement of gut microbiota in the development of low-grade inflammation and type 2 diabetes associated with obesity. *Gut Microbes, 3*(4), 279–288. http://doi.org/10.4161/gmic.19625

Cho, I., & Blaser, M. J. (2012). The human microbiome: At the interface of health and disease. *Nature Reviews Genetics, 13*. http://doi.org/10.1038/nrg3182

Corr, S. C., Li, Y., Riedel, C. U., O'Toole, P. W., Hill, C., & Gahan, C. G. M. (2007). Bacteriocin production as a mechanism for the antiinfective activity of *Lactobacillus salivarius* UCC118. *Proceedings of the National Academy of Sciences, 104*(18), 7617–7621. http://doi.org/10.1073/ pnas.0700440104

Dang, Y., Reinhardt, J. D., Zhou, X., & Zhang, G. (2014). The effect of probiotics supplementation on helicobacter pylori eradication rates and side effects during eradication therapy: A meta-analysis. *PLoS ONE, 9*(11), e111030. http://doi.org/10.1371/journal.pone. 0111030

De Paepe, M., Leclerc, M., Tinsley, C. R., & Petit, M.-A. (2014). Bacteriophages: An underestimated role in human and animal health? *Frontiers in Cellular and Infection Microbiology, 4*, 39. http://doi.org/10.3389/fcimb.2014.00039

Delzenne, N. M., & Cani, P. D. (2011). Interaction between obesity and the gut microbiota: Relevance in nutrition. *Annual Review of Nutrition, 31*, 15–31. http://doi.org/10.1146/annurev-nutr-072610-145146

Dimidi, E., Christodoulides, S., Fragkos, K. C., Scott, S. M., & Whelan, K. (2014). The effect of probiotics on functional constipation in adults: a systematic review and meta-analysis of randomized controlled trials. *The American Journal of Clinical Nutrition, 100*(4), 1075–1084. http://doi.org/10.3945/ajcn.114.089151

Ding, S., Chi, M. M., Scull, B. P., Rigby, R., Schwerbrock, N. M. J., Magness, S., Lund, P. K., et al. (2010). High-fat diet: Bacteria interactions promote intestinal inflammation which precedes and correlates with obesity and insulin resistance in mouse. *PloS One, 5*(8), e12191. http://doi.org/10.1371/journal.pone.0012191

Doege, K., Grajecki, D., Zyriax, B.-C., Detinkina, E., Zu Eulenburg, C., & Buhling, K. J. (2012). Impact of maternal supplementation with probiotics during pregnancy on atopic eczema in childhood–a meta-analysis. *The British Journal of Nutrition, 107*(1), 1–6. http://doi.org/10. 1017/S0007114511003400

Duerkop, B. A., Clements, C. V., Rollins, D., Rodrigues, J. L. M., & Hooper, L. V. (2012). A composite bacteriophage alters colonization by an intestinal commensal bacterium. *Proceedings of the National Academy of Sciences, 5*(43), 1–11. http://doi.org/10.1073/pnas. 1206136109

Dutilh, B. E., Cassman, N., McNair, K., Sanchez, S. E., Silva, G. G. Z., Boling, L., Edwards, R. A., et al. (2014). A highly abundant bacteriophage discovered in the unknown sequences of human faecal metagenomes. *Nature Communications, 5*, 1–11. http://doi.org/10.1038/ ncomms5498

Edgar, R., Friedman, N., Shahar, M. M., & Qimron, U. (2012). Reversing bacterial resistance to antibiotics by phage-mediated delivery of dominant sensitive genes. *Applied and Environmental Microbiology, 78*(3), 744–751. http://doi.org/10.1128/AEM.05741-11

Eriksson, F., Tsagozis, P., Lundberg, K., Parsa, R., Mangsbo, S. M., Persson, M. A. A., Pisa, P., et al. (2009). Tumor-specific bacteriophages induce tumor destruction through activation of tumor-associated macrophages. *Journal of Immunology, 182*(5), 3105–3111. http://doi.org/10. 4049/jimmunol.0800224

Fiorentin, L. (2005). Oral treatment with bacteriophages reduces the concentration of Salmonella Enteritidis PT4 in caecal contents of broilers. *Avian Pathology, 34*(3), 258–263. Retrieved from http://resolver.scholarsportal.info/resolve/03079457/v34i0003/258_otwbrtpiccob.xml

Flint, H. J., Bayer, E. A., Rincon, M. T., Lamed, R., & White, B. A. (2008). Polysaccharide utilization by gut bacteria: Potential for new insights from genomic analysis. *Nature Reviews Microbiology, 6*(2), 121–131. http://doi.org/10.1038/nrmicro1817

Ghosh, D., Roy, K., Williamson, K. E., Srinivasiah, S., Wommack, K. E., & Radosevich, M. (2009). Acyl-homoserine lactones can induce virus production in lysogenic bacteria: An alternative paradigm for prophage induction. *Applied and Environmental Microbiology, 75*(22), 7142–7152. http://doi.org/10.1128/AEM.00950-09

Gill, J. J., & Young, R. (2011). Therapeutic applications of phage biology: History, practice, and recommendations. In *Emerging trends in antibacterial discovery: Answering the call to arms* (pp. 367–410). Horizon Scientific Press. Retrieved from https://books.google.com/books?hl= en&lr=&id=B_GhBK7sgWIC&pgis=1

Goldman, P., Peppercorn, M. A., & Goldin, B. R. (1974). Metabolism of drugs by microorganisms in the intestine. *The American Journal of Clinical Nutrition, 27*, 1348–1355.

Grangette, C., Nutten, S., Palumbo, E., Morath, S., Hermann, C., Dewulf, J., Mercenier, A., et al. (2005). Enhanced antiinflammatory capacity of a *Lactobacillus plantarum* mutant synthesizing modified teichoic acids. *Proceedings of the National Academy of Sciences, 102*, 10321–10326. http://doi.org/10.1073/pnas.0504084102

Greer, G. G. (2005). Bacteriophage control of foodborne bacteriat. *Journal of Food Protection, 68*(5), 1102–1111.

Hagens, S., & Bläsi, U. (2003). Genetically modified filamentous phage as bactericidal agents: A pilot study. *Letters in Applied Microbiology, 37*, 318–323. http://doi.org/10.1046/j.1472-765X. 2003.01400.x

Hagens, S., Habel, A., von Ahsen, U., von Gabain, A., & Blasi, U. (2004). Therapy of experimental Pseudomonas infections with a nonreplicating genetically modified phage. *Antimicrobial Agents and Chemotherapy, 48*(10), 3817–3822. http://doi.org/10.1128/AAC.48. 10.3817-3822.2004

Hamer, H. M., Jonkers, D., Venema, K., Vanhoutvin, S., Troost, F. J., & Brummer, R. J. (2008). Review article: The role of butyrate on colonic function. *Alimentary Pharmacology and Therapeutics, 27*(2), 104–119. http://doi.org/10.1111/j.1365-2036.2007.03562.x

Hao, Q., Lu, Z. Z., Dong, B. R., Huang, C. Q., & Wu, T. (2011). Probiotics for preventing acute upper respiratory tract infections. *Cochrane Database of Systematic Reviews, 9*(9), CD006895. http://doi.org/10.1002/14651858.CD006895.pub2

Hargreaves, K. R., Kropinski, A. M., & Clokie, M. R. J. (2014). What does the talking? Quorum sensing signalling genes discovered in a bacteriophage genome. *PLoS ONE, 9*(1), e85131. http://doi.org/10.1371/journal.pone.0085131

Heilmann, S., Sneppen, K., & Krishna, S. (2012). Coexistence of phage and bacteria on the boundary of self-organized refuges. *Proceedings of the National Academy of Sciences, 109* (31), 12828–12833. http://doi.org/10.1073/pnas.1200771109

Hempel, S., Newberry, S. J., Maher, A. R., Wang, Z., Miles, J. N. V, Shanman, R., Shekelle, P. G., et al. (2012). Probiotics for the prevention and treatment of antibiotic-associated diarrhea. *JAMA, 307*(18), 1959–1969.

Hill, D. A., & Artis, D. (2010). Intestinal bacteria and the regulation of immune cell homeostasis. *Annual Review of Immunology, 28*, 623–667. http://doi.org/10.1146/annurev-immunol-030409-101330

Hill, C., Guarner, F., Reid, G., Gibson, G. R., Merenstein, D. J., Pot, B., Sanders, M. E., et al. (2014). Expert consensus document: The International Scientific Association for Probiotics and Prebiotics consensus statement on the scope and appropriate use of the term probiotic. *Nature Reviews Gastroenterology & Hepatology, 11*(8), 9. http://doi.org/10.1038/nrgastro.2014.66

Holte, K., Krag, A., & Gluud, L. L. (2012). Systematic review and meta-analysis of randomized trials on probiotics for hepatic encephalopathy. *Hepatology Research, 42*(10), 1008–1015. http://doi.org/10.1111/j.1872-034X.2012.01015.x

Hooper, L. V, Wong, M. H., Thelin, A., Hansson, L., Falk, P. G., & Gordon, J. I. (2001). Molecular analysis of commensal host-microbial relationships in the intestine. *Science, 291* (5505), 881–884. http://doi.org/10.1126/science.291.5505.881

Hornef, M. W., & Bogdan, C. (2005). The role of epithelial Toll-like receptor expression in host defense and microbial tolerance. *Journal of Endotoxin Research, 11*(2), 124–128. http://doi. org/10.1177/09680519050110020901

Hoveyda, N., Heneghan, C., Mahtani, K. R., Perera, R., Roberts, N., & Glasziou, P. (2009). A systematic review and meta-analysis: Probiotics in the treatment of irritable bowel syndrome. *BMC Gastroenterology, 9*, 15. http://doi.org/10.1186/1471-230X-9-15

Hungin, a. P. S., Mulligan, C., Pot, B., Whorwell, P., Agréus, L., Fracasso, P., De Wit, N., et al. (2013). Systematic review: Probiotics in the management of lower gastrointestinal symptoms in clinical practice—An evidence-based international guide. *Alimentary Pharmacology and Therapeutics, 38*(8), 864–886. http://doi.org/10.1111/apt.12460

Huttenhower, C., Gevers, D., Knight, R., Abubucker, S., Badger, J. H., Chinwalla, A. T., White, O., et al. (2012). Structure, function and diversity of the healthy human microbiome. *Nature, 486*(7402), 207–214. http://doi.org/10.1038/nature11234

Izcue, A., Coombes, J. L., & Powrie, F. (2006). Regulatory T cells suppress systemic and mucosal immune activation to control intestinal inflammation. *Immunological Reviews, 212*, 256–271. http://doi.org/10.1111/j.0105-2896.2006.00423.x

JAMA. (2014). Probiotics revisited. *JAMA, 312*(17), 1796. http://doi.org/10.1001/jama.2014. 14389

Jamalludeen, N., Johnson, R. P., Shewen, P. E., & Gyles, C. L. (2009). Evaluation of bacteriophages for prevention and treatment of diarrhea due to experimental enterotoxigenic *Escherichia coli* O149 infection of pigs. *Veterinary Microbiology, 136*(1–2), 135–141. http:// doi.org/10.1016/j.vetmic.2008.10.021

Jayaraman, A., & Wood, T. K. (2008). Bacterial quorum sensing: Signals, circuits, and implications for biofilms and disease. *Annual Review of Biomedical Engineering, 10*, 145–167. http://doi.org/10.1146/annurev.bioeng.10.061807.160536

Johnston, B. C., Ma, S. S. Y., Goldenberg, J. Z., Thorlund, K., Vandvik, P. O., Loeb, M., Guyatt, G. H., et al. (2012). Probiotics for the prevention of clostridium difficile–associated diarrhea: A systematic review and meta-analysis. *Annals of Internal Medicine, 157*, 878–888. Retrieved from http://annals.org/article.aspx?articleid=1390418

Joyce, S. A., MacSharry, J., Casey, P. G., Kinsella, M., Murphy, E. F., Shanahan, F., Gahan, C. G. M., et al. (2014). Regulation of host weight gain and lipid metabolism by bacterial bile acid modification in the gut. *Proceedings of the National Academy of Sciences, 111*(20), 7421–7426. http://doi.org/10.1073/pnas.1323599111

Kalliomäki, M., Salminen, S., Poussa, T., Arvilommi, H., & Isolauri, E. (2003). Probiotics and prevention of atopic disease: 4-year follow-up of a randomised placebo-controlled trial. *Lancet, 361*, 1869–1871. http://doi.org/10.1016/S0140-6736(03)13490-3

Keen, E. C. (2015). A century of phage research: Bacteriophages and the shaping of modern biology. *BioEssays, 37*(1), 6–9. http://doi.org/10.1002/bies.201400152

Kim, M.-S., Park, E.-J., Roh, S. W., & Bae, J.-W. (2011). Diversity and abundance of single-stranded DNA viruses in human feces. *Applied and Environmental Microbiology, 77*(22), 8062–8070. http://doi.org/10.1128/AEM.06331-11

Kranich, J., Maslowski, K. M., & Mackay, C. R. (2011). Commensal flora and the regulation of inflammatory and autoimmune responses. *Seminars in Immunology, 23*(2), 139–145. http://doi.org/10.1016/j.smim.2011.01.011

Krom, R. J., Bhargava, P., Lobritz, M. A., & Collins, J. J. (2015). Engineered phagemids for nonlytic, targeted antibacterial therapies. *Nano Letters, 15*(7), 4808–4813. http://doi.org/10.1021/acs.nanolett.5b01943

Lee, Y. K., & Mazmanian, S. K. (2010). Has the microbiota played a critical role in the evolution of the adaptive immune system? *Science, 330*(6012), 1768–1773. http://doi.org/10.1126/science.1195568

Lepper, P., Held, T., Schneider, E., Bölke, E., Gerlach, H., & Trautmann, M. (2002). Clinical implications of antibiotic-induced endotoxin release in septic shock. *Intensive Care Medicine, 28*(7), 824–833. http://doi.org/10.1007/s00134-002-1330-6

Libis, V. K., Bernheim, A. G., Basier, C., Jaramillo-Riveri, S., Deyell, M., Aghoghogbe, I., Wintermute, E. H., et al. (2014). Silencing of antibiotic resistance in *E. coli* with engineered phage bearing small regulatory RNAs. *ACS Synthetic Biology, 3*(12), 1003–1006. http://doi.org/10.1021/sb500033d

Louis, P., Hold, G. L., & Flint, H. J. (2014). The gut microbiota, bacterial metabolites and colorectal cancer. *Nature Reviews Microbiology, 12*(10), 661–672. http://doi.org/10.1038/nrmicro3344

Lu, T. K., & Collins, J. J. (2007). Dispersing biofilms with engineered enzymatic bacteriophage. *Proceedings of the National Academy of Sciences, 104*(27), 11197–11202. http://doi.org/10.1073/pnas.0704624104

Lu, T. K., & Collins, J. J. (2009). Engineered bacteriophage targeting gene networks as adjuvants for antibiotic therapy. *Proceedings of the National Academy of Sciences of the United States of America, 106*(12), 4629–4634. http://doi.org/10.1073/pnas.0800442106

Macfarlane, S., & Dillon, J. F. (2007). Microbial biofilms in the human gastrointestinal tract. *Journal of Applied Microbiology, 102*(5), 1187–1196. http://doi.org/10.1111/j.1365-2672. 2007.03287.x

Mack, D. R., Michail, S., Wei, S., McDougall, L., & Hollingsworth, M. A. (1999). Probiotics inhibit enteropathogenic *E. coli* adherence in vitro by inducing intestinal mucin gene expression. *American Journal of Physiology-Gastrointestinal and Liver Physiology, 276*(4), G941–G950. Retrieved from http://ajpgi.physiology.org/content/276/4/G941

Macpherson, A. J., & Harris, N. L. (2004). Interactions between commensal intestinal bacteria and the immune system. *Nature Reviews Immunology, 4*(6), 478–485. http://doi.org/10.1038/nri1373

Macpherson, A. J., Hunziker, L., McCoy, K., & Lamarre, A. (2001). IgA responses in the intestinal mucosa against pathogenic and non-pathogenic microorganisms. *Microbes and Infection, 3*(12), 1021–1035. http://doi.org/10.1016/S1286-4579(01)01460-5

Macpherson, A. J., & Uhr, T. (2004). Induction of protective IgA by intestinal dendritic cells carrying commensal bacteria. *Science, 303*, 1662–1665. http://doi.org/10.1126/science. 1091334

Makarova, K. S., Haft, D. H., Barrangou, R., Brouns, S. J. J., Charpentier, E., Horvath, P., Koonin, E. V., et al. (2011). Evolution and classification of the CRISPR-Cas systems. *Nature Reviews Microbiology, 9*(6), 467–477. http://doi.org/10.1038/nrmicro2577

Marchesi, J. R., & Shanahan, F. (2007). The normal intestinal microbiota. *Current Opinion in Infectious Disease, 20*(5), 508–513.

Martin, R., Nauta, A. J., Ben Amor, K., Knippels, L. M. J., Knol, J., & Garssen, J. (2010). Early life: Gut microbiota and immune development in infancy. *Beneficial Microbes, 1*(4), 367–382. http://doi.org/10.3920/BM2010.0027

Martínez, I., Stegen, J. C., Maldonado-Gómez, M. X., Eren, A. M., Siba, P. M., Greenhill, A. R., Walter, J., et al. (2015). The gut microbiota of rural papua new guineans: Composition, diversity patterns, and ecological processes. *Cell Reports, 11*(4), 527–538. http://doi.org/10.1016/j.celrep.2015.03.049

Mazmanian, S. K., Round, J. L., & Kasper, D. L. (2008). A microbial symbiosis factor prevents intestinal inflammatory disease. *Nature, 453*(7195), 620–625. http://doi.org/10.1038/nature07008

McFarland, L. V. (2007). Meta-analysis of probiotics for the prevention of traveler's diarrhea. *Travel Medicine and Infectious Disease, 5*(2), 97–105. http://doi.org/10.1016/j.tmaid.2005.10. 003

Merril, C. R., Biswas, B., Carltont, R., Jensen, N. C., Creed, G. J., Zullo, S., et al. (1996). Long-circulating bacteriophage as antibacterial agents. *Proceedings of the National Academy of Sciences, 93*, 3188–3192.

Miller, M. B., & Bassler, B. L. (2001). Quorum sensing in bacteria. *Annual Reviews in Microbiology, 55*, 165–199. Retrieved from http://www.annualreviews.org/doi/pdf/10.1146/annurev.micro.55.1.165

Miller, R. W., Skinner, J., Sulakvelidze, A., Mathis, G. F., & Hofacre, C. L. (2010). Bacteriophage therapy for control of necrotic enteritis of broiler chickens experimentally infected with Clostridium perfringens. *Avian Diseases, 54*(1), 33–40. Retrieved from http://www. aaapjournals.info/doi/abs/10.1637/8953-060509-Reg.1

Mills, S., Shanahan, F., Stanton, C., Hill, C., Coffey, A., & Ross, R. P. (2013). Movers and shakers: Influence of bacteriophages in shaping the mammalian gut microbiota. *Gut Microbes, 4*(1), 4–16. http://doi.org/10.4161/gutm.22371

Minot, S., Sinha, R., Chen, J., Li, H., Keilbaugh, S. A., Wu, G. D., Bushman, F. D., et al. (2011). The human gut virome: Inter-individual variation and dynamic response to diet. *Genome Research, 21*, 1616–1625. http://doi.org/10.1101/gr.122705.111

Modi, S. R., Collins, J. J., & Relman, D. A. (2014). Antibiotics and the gut microbiota. *The Journal of Clinical Investigation, 124*(10), 4212–4218. http://doi.org/10.1172/JCI72333

Modi, S. R., Lee, H. H., Spina, C. S., & Collins, J. J. (2013). Antibiotic treatment expands the resistance reservoir and ecological network of the phage metagenome. *Nature, 499*(7457), 219–222. http://doi.org/10.1038/nature12212

Nakonieczna, A., Cooper, C. J., & Gryko, R. (2015). Bacteriophages and bacteriophage derived endolysins as potential therapeutics to combat gram positive spore forming bacteria. *Journal of Applied Microbiology*, n/a–n/a. http://doi.org/10.1111/jam.12881

Norris, J. S., Westwater, C., & Schofield, D. (2000). Prokaryotic gene therapy to combat multidrug resistant bacterial infection. *Gene Therapy, 7*(9), 723–725. http://doi.org/10.1038/sj.gt.3301178

Oliveira, A., Sereno, R., Nicolau, A., & Azeredo, J. (2009). The influence of the mode of administration in the dissemination of three coliphages in chickens. *Poultry Science, 88*(4), 728–733. http://doi.org/10.3382/ps.2008-00378

Paul, V. D., Sundarrajan, S., Rajagopalan, S. S., Hariharan, S., Kempashanaiah, N., Padmanabhan, S., Ramachandran, J., et al. (2011). Lysis-deficient phages as novel therapeutic agents for controlling bacterial infection. *BMC Microbiology, 11*(1), 195. http://doi.org/10.1186/1471-2180-11-195

Pelucchi, C., Chatenoud, L., Turati, F., Galeone, C., Moja, L., Bach, J.-F., La Vecchia, C., et al. (2012). Probiotics supplementation during pregnancy or infancy for the prevention of atopic dermatitis: A meta-analysis. *Epidemiology, 23*(3), 402–414. http://doi.org/10.1097/EDE.0b013e31824d5da2

Petrof, E. O., Dhaliwal, R., Manzanares, W., Johnstone, J., Cook, D., & Heyland, D. K. (2012). Probiotics in the critically ill: A systematic review of the randomized trial evidence. *Critical Care Medicine, 40*(12), 3290–3302. http://doi.org/10.1097/CCM.0b013e318260cc33

Pullan, R. D., Thomas, G. A., Rhodes, M., Newcombe, R. G., Williams, G. T., Allen, A., Rhodes, J., et al. (1994). Thickness of adherent mucus gel on colonic mucosa in humans and its relevance to colitis. *Gut, 35*(3), 353–359. Retrieved from http://www.pubmedcentral.nih.gov/articlerender.fcgi?artid=1374589&tool=pmcentrez&rendertype=abstract

Qin, J., Li, R., Raes, J., Arumugam, M., Burgdorf, K. S., Manichanh, C., Wang, J., et al. (2010). A human gut microbial gene catalogue established by metagenomic sequencing. *Nature, 464* (7285), 59–65. http://doi.org/10.1038/nature08821

Qin, J., Li, Y., Cai, Z., Li, S., Zhu, J., Zhang, F., Wang, J., et al. (2012). A metagenome-wide association study of gut microbiota in type 2 diabetes. *Nature, 490*(7418), 55–60. http://doi.org/10.1038/nature11450

Rea, M. C., Alemayehu, D., Ross, R. P., & Hill, C. (2013). Gut solutions to a gut problem: Bacteriocins, probiotics and bacteriophage for control of Clostridium difficile infection. *Journal of Medical Microbiology, 62*(9), 1369–1378. http://doi.org/10.1099/jmm.0.058933-0

Rey, F. E., Faith, J. J., Bain, J., Muehlbauer, M. J., Stevens, R. D., Newgard, C. B., Gordon, J. I., et al. (2010). Dissecting the in vivo metabolic potential of two human gut acetogens. *The Journal of Biological Chemistry, 285*(29), 22082–22090. http://doi.org/10.1074/jbc.M110.117713

Reyes, A., Haynes, M., Hanson, N., Angly, F. E., Heath, A. C., Rohwer, F., Gordon, J. I., et al. (2010). Viruses in the faecal microbiota of monozygotic twins and their mothers. *Nature, 466* (7304), 334–338. http://doi.org/10.1038/nature09199

Reyes, A., Semenkovich, N. P., Whiteson, K., Rohwer, F., & Gordon, J. I. (2012). Going viral: Next-generation sequencing applied to phage populations in the human gut. *Nature Reviews. Microbiology, 10*(9), 607–617. http://doi.org/10.1038/nrmicro2853

Ritchie, M. L., & Romanuk, T. N. (2012). A meta-analysis of probiotic efficacy for gastrointestinal diseases. *PloS ONE, 7*(4), e34938. http://doi.org/10.1371/journal.pone.0034938

Rodriguez-Valera, F., Martin-Cuadrado, A.-B., Rodriguez-Brito, B., Pasić, L., Thingstad, T. F., Rohwer, F., & Mira, A. (2009). Explaining microbial population genomics through phage predation. *Nature Reviews Microbiology, 7*(11), 828–836. http://doi.org/10.1038/nrmicro2235

Sazawal, S., Hiremath, G., Dhingra, U., Malik, P., Deb, S., & Black, R. E. (2006). Efficacy of probiotics in prevention of acute diarrhoea: A meta-analysis of masked, randomised, placebo-controlled trials. *The Lancet Infectious Diseases, 6*, 374–382. http://doi.org/10.1016/S1473-3099(06)70495-9

Seed, K. D., Yen, M., Shapiro, B. J., Hilaire, I. J., Charles, R. C., Teng, J. E., Camilli, A., et al. (2014). Evolutionary consequences of intra-patient phage predation on microbial populations. *eLife, 3*, e03497. http://doi.org/10.7554/eLife.03497

Selva, L., Viana, D., Regev-Yochay, G., Trzcinski, K., Corpa, J. M., Lasa, I., Penadés, J. R., et al. (2009). Killing niche competitors by remote-control bacteriophage induction. *Proceedings of the National Academy of Sciences, 106*(4), 1234–1238. http://doi.org/10.1073/pnas.0809600106

Senok, A. C., Ismaeel, A. Y., & Botta, G. A. (2005). Probiotics: Facts and myths. *Clinical Microbiology and Infection, 11*(12), 958–966. http://doi.org/10.1111/j.1469-0691.2005.01228.x

Shanahan, F. (2012). The microbiota in inflammatory bowel disease: Friend, bystander, and sometime-villain. *Nutrition Reviews, 70*(Suppl. 1), 31–37. http://doi.org/10.1111/j.1753-4887.2012.00502.x

Sjögren, K., Engdahl, C., Henning, P., Lerner, U. H., Tremaroli, V., Lagerquist, M. K., Ohlsson, C., et al. (2012). The gut microbiota regulates bone mass in mice. *Journal of Bone and Mineral Research, 27*(6), 1357–1367. http://doi.org/10.1002/jbmr.1588

Smith, H. W., Huggins, M. B., & Shaw, K. M. (1987). The control of experimental *Escherichia coli* diarrhoea in calves by means of bacteriophages. *Journal of General Microbiology, 133*(5), 1111–1126.

Smith, R. S., & Harris, S. G. (2002). The pseudomonas aeruginosa quorum-sensing molecule N-(3-Oxododecanoyl) Homoserine Lactone contributes to virulence and induces inflammation in vivo. *Journal of Bacteriology, 184*(4), 1132–1139. http://doi.org/10.1128/JB.184.4.1132–1139.2002

Smits, H. H., Engering, A., van der Kleij, D., de Jong, E. C., Schipper, K., van Capel, T. M. M., Kapsenberg, M. L., et al. (2005). Selective probiotic bacteria induce IL-10-producing regulatory T cells in vitro by modulating dendritic cell function through dendritic cell-specific intercellular adhesion molecule 3-grabbing nonintegrin. *The Journal of Allergy and Clinical Immunology, 115*(6), 1260–1267. http://doi.org/10.1016/j.jaci.2005.03.036

Sommer, F., & Bäckhed, F. (2013). The gut microbiota–masters of host development and physiology. *Nature Reviews Microbiology, 11*(4), 227–238. http://doi.org/10.1038/nrmicro2974

Stappenbeck, T. S., Hooper, L. V, & Gordon, J. I. (2002). Developmental regulation of intestinal angiogenesis by indigenous microbes via Paneth cells. *Proceedings of the National Academy of Sciences, 99*(24), 15451–15455. http://doi.org/10.1073/pnas.202604299

Stecher, B., Maier, L., & Hardt, W.-D. (2013). "Blooming" in the gut: How dysbiosis might contribute to pathogen evolution. *Nature Reviews Microbiology, 11*(4), 277–284. http://doi.org/10.1038/nrmicro2989

Sulakvelidze, A. (2011). The challenges of bacteriophage therapy. *Industrial Pharmacy, 45*(31), 14–18. http://doi.org/10.1128/AAC.45.3.649

Szajewska, H., Ruszczyński, M., & Radzikowski, A. (2006). Probiotics in the prevention of antibiotic-associated diarrhea in children: a meta-analysis of randomized controlled trials. *The Journal of Pediatrics, 149*(3), 367–372. http://doi.org/10.1016/j.jpeds.2006.04.053

Tamboli, C. P., Neut, C., Desreumaux, P., & Colombel, J. F. (2004). Dysbiosis in inflammatory bowel disease. *Gut, 53*, 1–4. http://doi.org/10.1136/gut.2007.134668

Thiel, K. (2004). Old dogma, new tricks—21st century phage therapy. *Nature Biotechnology, 22*(1), 31–36. http://doi.org/10.1038/nbt0104-31

Thingstad, T. F. (2000). Elements of a theory for the mechanisms controlling abundance, diversity, and biogeochemical role of lytic bacterial viruses in aquatic systems. *Limnology and Oceanography, 45*(6), 1320–1328. http://doi.org/10.4319/lo.2000.45.6.1320

Tsonos, J., Vandenheuvel, D., Briers, Y., De Greve, H., Hernalsteens, J.-P., & Lavigne, R. (2014). Hurdles in bacteriophage therapy: Deconstructing the parameters. *Veterinary Microbiology, 171*(3–4), 460–469. http://doi.org/10.1016/j.vetmic.2013.11.001

Turnbaugh, P. J., Hamady, M., Yatsunenko, T., Cantarel, B. L., Duncan, A., Ley, R. E., Gordon, J. I., et al. (2009). A core gut microbiome in obese and lean twins. *Nature, 457*(7228), 480–484. http://doi.org/10.1038/nature07540

Turnbaugh, P. J., Ley, R. E., Hamady, M., Fraser-Liggett, C. M., Knight, R., & Gordon, J. I. (2007). The human microbiome project. *Nature, 449*(7164), 804–810. http://doi.org/10.1038/nature06244

Turnbaugh, P. J., Ridaura, V. K., Faith, J. J., Rey, F. E., Knight, R., & Gordon, J. I. (2009). The effect of diet on the human gut microbiome: A metagenomic analysis in humanized gnotobiotic mice. *Science Translational Medicine, 1*(6), 6ra14. http://doi.org/10.1126/scitranslmed.3000322

Vanderhaeghen, S., Lacroix, C., & Schwab, C. (2015). Methanogen communities in stools of humans of different age and health status and co-occurrence with bacteria. *FEMS Microbiology Letters, 362*(13), fnv092. http://doi.org/10.1093/femsle/fnv092

Ventura, M., Sozzi, T., Turroni, F., Matteuzzi, D., & van Sinderen, D. (2011). The impact of bacteriophages on probiotic bacteria and gut microbiota diversity. *Genes & Nutrition, 6*(3), 205–207. http://doi.org/10.1007/s12263-010-0188-4

Videlock, E. J., & Cremonini, F. (2012). Meta-analysis: Probiotics in antibiotic-associated diarrhoea. *Alimentary Pharmacology and Therapeutics, 35*, 1355–1369. http://doi.org/10.1111/j.1365-2036.2012.05104.x

Wassenaar, T. M., & Panigrahi, P. (2014). Is a foetus developing in a sterile environment? *Letters in Applied Microbiology, 59*(6), 572–579. http://doi.org/10.1111/lam.12334

Wehkamp, J., Harder, J., Wehkamp, K., Wehkamp-von Meissner, B., Schlee, M., Enders, C., Stange, E. F., et al. (2004). NF-kappaB- and AP-1-mediated induction of human beta defensin-2 in intestinal epithelial cells by *Escherichia coli* Nissle 1917: A novel effect of a probiotic bacterium. *Infection and Immunity, 72*(10), 5750–5758. http://doi.org/10.1128/IAI.72.10.5750-5758.2004

Westwater, C., Kasman, L. M., Schofield, D. A., Werner, P. A., Dolan, J. W., Schmidt, M. G., Norris, J. S., et al. (2003). Use of genetically engineered phage to deliver antimicrobial agents to bacteria: An alternative therapy for treatment of bacterial infections. *Antimicrobial Agents and Chemotherapy, 47*(4), 1301–1307. http://doi.org/10.1128/AAC.47.4.1301-1307.2003

Wiggins, B. A., & Alexander, M. (1985). Minimum bacterial density for bacteriophage replication: implications for significance of bacteriophages in natural ecosystems. *Applied and Environmental Microbiology, 49*(1), 19–23.

Wolin, M. J. (1981). Fermentation in the rumen and human large intestine. *Science, 213*(4515), 1463–1468. http://doi.org/10.1126/science.7280665

Yacoby, I., Bar, H., & Benhar, I. (2007). Targeted drug-carrying bacteriophages as antibacterial nanomedicines. *Antimicrobial Agents and Chemotherapy, 51*(6), 2156–2163. http://doi.org/10.1128/AAC.00163-07

Yacoby, I., & Benhar, I. (2008). Targeted filamentous bacteriophages as therapeutic agents. *Expert Opinion on Drug Delivery, 5*(3), 321–329. http://doi.org/10.1517/17425247.5.3.321

Yacoby, I., Shamis, M., Bar, H., Shabat, D., & Benhar, I. (2006). Targeting antibacterial agents by using drug-carrying filamentous bacteriophages. *Antimicrobial Agents and Chemotherapy, 50*(6), 2087–2097. http://doi.org/10.1128/AAC.00169-06

Yan, F., Cao, H., Cover, T. L., Whitehead, R., Washington, M. K., & Polk, D. B. (2007). Soluble proteins produced by probiotic bacteria regulate intestinal epithelial cell survival and growth. *Gastroenterology, 132*(2), 562–575. http://doi.org/10.1053/j.gastro.2006.11.022

Yan, F., & Polk, D. B. (2002). Probiotic bacterium prevents cytokine-induced apoptosis in intestinal epithelial cells. *Journal of Biological Chemistry, 277*(52), 50959–50965. http://doi.org/10.1074/jbc.M207050200

Yatsunenko, T., Rey, F. E., Manary, M. J., Trehan, I., Dominguez-Bello, M. G., Contreras, M., Gordon, J. I., et al. (2012). Human gut microbiome viewed across age and geography. *Nature, 482*, 331–338. http://doi.org/10.1038/nature11053

Young, R. (2002). Bacteriophage holins: Deadly diversity. *Journal of Molecular Microbiology and Biotechnology, 4*(1), 21–36.

Young, V. B., Kahn, S. A., Schmidt, T. M., & Chang, E. B. (2011). Studying the enteric microbiome in inflammatory bowel diseases: Getting through the growing pains and moving forward. *Frontiers in Microbiology, 2*, 144. http://doi.org/10.3389/fmicb.2011.00144

Chapter 6
Phage for Biodetection

Abstract Bacterial threats pose a major global health issue and are the culprits behind extensive morbidity and mortality each year. Many of these biological entities have circumvented our ability to detect them via standard biological testing methods, where earliest detection of a biological threat may occur at the time of clinical diagnosis. These standard methods typically utilize biochemical and genotypic tests which, while moderately effective, can be expensive, laborious, time consuming and offer varying levels of success. Furthermore, by the time of detection, particularly with respect to biological entities relevant to of bioterrorism and environmental contamination, it may be too late. As such, there is a great demand for new bacterial detection technologies, predominantly in the: (i) medical industry where earlier diagnosis of infectious diseases would confer better therapeutic outcomes; (ii) agricultural, food and water processing industries to benefit from the ability to identify disease-causing organism(s) in order to prevent and control ongoing outbreaks; and (iii) the military and/or academic industries to identify and prevent agents of bioterrorism (Gulig in Principles of bacterial detection: biosensors, recognition receptors and microsystems (Springer, Montreal, pp. 755–783, 2008). New and emerging non-bacteriophage-based techniques to detect pathogens include both those for nucleic acid detection (i.e. real-time PCR, isothermal-PCR and microarrays) and biosensors (i.e. ELISA, antibody arrays, fiber optics and surface plasmon resonance), each of which has been reviewed extensively (Gopinath et al in Biosens Bioelectron 60: 332–342, 2014; Gulig in Principles of bacterial detection: biosensors, recognition receptors and microsystems. Springer, Montreal, pp. 755–783, 2008). Bacteriophages possess the innate ability to specifically target and amplify in a bacterial host—key attributes for effective bacterial detection (Singh et al. in Analyst 137: 3405, 2012). In this chapter we will examine these phage-based bacterial sensor strategies.

© The Author(s) 2016
J. Nicastro et al., *Bacteriophage Applications—Historical Perspective and Future Potential*, SpringerBriefs in Biochemistry and Molecular Biology, DOI 10.1007/978-3-319-45791-8_6

1 Introduction to Phage-Based Biodetection

Bacteriophages are intracellular bacterial parasites that bind specific host surface receptors and utilize the bacterial cell machinery for their own multiplication and dissemination of mature virions (Housby and Mann 2009). Phages possess almost infinite potential for the manipulation of the bacterial species, including bacterial detection and phage-based detection technologies. While still currently in the infancy stages, these embrace modern techniques and advantages from modern biotechnology and nanotechnology (Singh et al. 2006). Phage offer many advantages that have already been observed and the outcomes from phage-based strategies have proven quicker, usually within days, in comparison to antisera/antibody methods, which can take up to months using conventional methods (Ulitzur and Ulitzur 2006). Phage typing has been used for decades and most extensively applied to the detection of *Mycobacterium, Escherichia, Pseudomonas, Salmonella* and *Campylobacter* species (Barry et al. 1996).

2 Plaque Assays

Phage infection of bacterial cells can be visualized on a petri plate as a clearing or "plaque" on an otherwise opaque bacterial lawn. Phage plaques, have been used traditionally as a means of bacterial detection and typing, whereby the phage infection and the viral amplification event can be visualized as a plaque forming unit (pfu). Plaque formation begins when susceptible bacterial species are infected by a phage of suitable host-range, resulting in the lysis of the host bacterial cell, followed by subsequent phage progeny release and infection of neighboring host cells. The result of these actions can be seen as plaques, which are normally visible under natural light and can be counted to give a positive indication of the pathogen (Cox 2012; Kalniņa et al. 2008). Traditional plaque assays have been used in the detection of pathogens, including but not limited to, *Bacillus anthracis* (Brigati et al. 2004; Gillis and Mahillon 2014; Thal and Nordberg 1968), *Staphylococcus aureus* (Wallmark et al. 1978), *Mycobacterium* species (Broxmeyer et al. 2002), *Escherichia coli* (Oda et al. 2004), *Pseudomonas* species (Brokopp et al. 1977); and *Campylobacter* species (Grajewski et al. 1985).

This traditional method is typically laborious and time consuming (Cox 2012). However, an advanced approach developed by Stewart et al. (1998) allows for a quicker and simpler method for phage-based assays cutting down processing time to 4 h in comparison to days. In this method samples are incubated with the specific phage and the exogenous phage excess is killed by a virucidal agent such as pomegranate rind extract. The susceptible bacterium can then be added to culture medium to enable plaque formation and grown from this point with a helper bacterium rather than pre-culturing suspected pathogens, dramatically cutting down the processing time (Stewart et al. 1998). This technique was successfully applied to

Mycobacterium tuberculosis, a slow growing bacterium, paired with a fast growing *Mycobacterium smegmatis* as a helper bacterium, cutting down the processing time from multiple weeks to two days (Minion et al. 2010; Rishi et al. 2014).

2.1 Phage Display for the Improvement of Plaque Assays

Bacteriophage specificity for bacterial cells enables them to be easily used for the typing of bacterial strains and pathogens; however, these methods possess a number of limitations that can lead to false negatives. In particular, most phage strains are specific to a small subset of their pathogenic hosts, where some of the pathogens may not be susceptible to the indicator phage causing a false negative. This natural specificity can be improved by expressing targeting peptides onto the capsid surface of the phage through a thoroughly developed process termed "phage display" (Nicastro et al. 2013; Smith and Petrenko 1997). The most commonly used peptide in this advancement is the single chain F variable (scFv) portion of an antibody that contains the antigen-binding regions of the heavy and light chains; the antibody portion is normally obtained from animals that are immunized with the antigen. Phage particles incorporating the fusion peptide of interest to their capsids are then selected by a process called "panning", which involves the display/binding of the antigen to the phage and the subsequent washing away of any unbound phage. The bound phage can then be eluted and amplified in the appropriate host cell (Smith and Petrenko 1997); Gulig et al. 2008). While the phage bound with scFv have been successful for the purposes of bacterial detection, the scFv section alone is not.

Sorokulova et al. (2005) developed a random 8-mer landscape phage library expressing fusions to coat protein (pVIII) of fd phage. The phage were panned against *S. typhimurium* whole cells resulting in the isolation of a phage that was highly specific towards *Salmonella* whole cells (Sorokulova et al. 2005). In a follow-up study, Olsen et al. (2006) were able to detect *Salmonella* with this phage at titers as low as 100 cells/mL (Olsen et al. 2006). This technique was also used to pan against *Bacillus anthracis*, an organism of particular relevance to bioterrorism, leading to the identification of multiple phage with the ability to capture anthrax spores with reasonable specificity (Brigati et al. 2004).

Ide et al. (2003) used the New England Biolabs 12-mer library of random peptides in an effort to isolate peptides for the identification of the H7 flagella of *E. coli* O157: H7 (Enterohemorrhagic *E. coli*, or EHEC). This is a pathogenic strain of bacteria that is responsible for the onset of hemorrhagic colitis and hemolytic uremic syndrome (HUS) due to the release of Shiga toxin (H7 antigen) that is shared by most strains and encoded by the resident prophage, 933W. They were able to develop specific clones to H7, one of which could bind to intact *E. coli* cells expressing flagella (Ide et al. 2003). Similarly, Turnbough (2003) developed phage libraries using New England Biolabs 7- and 12-mer libraries to isolate phage clones that could recognize a variety of *Bacillus* spores, including those formed by *B. anthracis* species. They were also able to differentiate between the spore types, an important consideration in the development of anti-spore reagents (Turnbough 2003).

The future of phage display biotechnology platforms seems promising with non-immunoglobulin target-binding tools at the forefront. These tools can be engineered to increase the randomness of binding. Examples include: (i) affibodies—small proteins isolated from *S. aureus* protein A that can be manipulated by randomizing the binding domain and highly specific bodies to protphageein targets, and have already been obtained; and (ii) anticalins—proteins based on lipocalin proteins that transport or store molecules that are soluble and can be manipulated to recognize a variety of targets with high affinity binding (Gulig et al. 2008). The ability of phage to amplify and form plaques can be quite effective in bacterial detection and has set the foundation for advancements to the field, including the use of phage as indicator organisms, addressed in the following section.

3 Bacteriophage Indicator Organisms (Reporter Phage)

It is highly unlikely that phage plaque assays, even in their optimized form, will become the ultimate tool in pathogen biological detection systems. Although plaque assays offer a cost-effective and reliable method for bacterial detection, there are many approaches to employ the natural abilities of the phages toward the development of more sophisticated detection strategies that can offer real-time, or near-real-time results. Bacteriophage indicator organisms can do just this by way of combining the natural binding and infection abilities of bacteriophage with a mode of detecting the phage itself. Bacteriophage indicator organisms, referred to as reporter phage, are based on the engineering of recombinant phage to transduce target bacterial hosts and express a reporter gene(s). Expression of the phage-encoded reporter gene(s) upon infection enables the subsequent identification of the infected host.

3.1 Fluorescence-Based Assays

In general, various methods for light emission microscopy permit simplistic microbial detection, and these can be used in combination with molecular cloning techniques to engineer phage to carry and deliver specific reporter genes such as luciferase or green florescent protein (GFP) that are expressed upon infection of the bacterial host. Among all the described reporter phage, those carrying a luciferase gene for bacterial detection account for the largest share. The bioluminescent signal generated by luciferase activity on luciferin is highly sensitive and rarely found in native food samples, thereby offering a high signal-to-noise ratio (Brovko et al. 2012; Schmelcher and Loessner 2014). Ulitzur and Kuhn (1987) were the first to construct a luciferase reporter phage (LRP). The group engineered the LRP by inserting a *luc* gene into a bacteriophage lambda (λ) cloning vector, thereby enabling the detection of an *E. coli* sample size as small as ten bacteria within an hour in milk (Ulitzer and Ulitzur 2006; Ulitzur and Kuhn 1987).

LRPs have now been developed for the detection of many pathogenic bacteria including: *E. coli*, *Salmonella species*, *Listeria species*, *S. aureus*, and *Mycobacterium species* (Loessner et al. 1996; Ripp et al. 2006; Tawil et al. 2014). Loessner et al. (1996) developed reporter phage A511::*luxAB*, which utilizes the broad host range of the *Listeria monocytogenes*-specific phage A511. The reporter phage strategy employed the phage display technique to conjugate a luciferase gene fusion immediately downstream of the major capsid protein gene of A511. The group reported emission data that supported the ability of the reporter phage to infect host cells in a manner such that the luciferase signal could be detected. The detection was highly sensitive, where positive indicators of infection were found even in the presence of low numbers of *Listeria species* bacteria within 24 h (Loessner et al. 1996).

Green Fluorescent Protein (GFP) has become one of the most frequently used biomarkers in molecular biology. In contrast to luciferase, GFP does not require a substrate (such as luciferin by luciferase) as the protein itself expresses a chromophore, thereby further simplifying fluorescent detection (Vesuna et al. 2005). Funatsu et al. (2002) were the first to use *GFP* reporter genes in a phage through direct cloning to develop a recombinant λ reporter phage. The group employed fluorescent microscopy to detect GFP in transduced bacteria within six hours post-infection (Loessner et al. 1997; Funatsu et al. 2002). In another study by Oda et al. (2004), phage display was used to display GFP on the surface of the capsid protein of the virulent phage PP01, specific to *E. coli* O157:H7. Fluorescent microscopy was used to visualize infected target cells and GFP signals were found as soon as 10 min post-infection. Interestingly, the investigators reported being able to detect a signal in viable cells, viable but non-culturable cells, and most surprisingly, even in dead cells; although the test was reported to be much less sensitive in the latter (Oda et al. 2004). In a follow-up study the group used T4 that was deficient in the ability to produce T4 lysozyme, enabling the phage to remain in the target cells without lysing the hosts. Detection and visualization of the targeted cells was relatively efficient as GFP continued to accumulate within (Tanji et al. 2004).

4   Immobilized Phage Particles as Probes for Bacterial Detection

Phage infection depends on the adsorption of the phage particle to the surface of its host. This occurs via specific recognition of the receptor molecules of the bacterial host cell, typically outer membrane proteins, flagella or pili (Berkane et al. 2006). Tailed phages, which make up the majority of the bacterial viruses, possess tail fibers that are responsible for the interaction with the bacterial receptors. Once this interaction occurs, adsorption is enabled and occurs in two stages: (i) reversible binding of the tail fiber to the cellular receptor(s); and (ii) the irreversible docking of the phage plate to the cell envelope. The specificity of the adsorption process is an ideal feature of bacteriophage that can be exploited for the detection of bacterial pathogens.

Bennett et al. (1997), were the first to report the use of immobilized phage, passively attached to a solid phase, to remove *Salmonella* as well as other species of *Enterobacteriaceae* from food materials. The group used the Felix 01 phage, originally found during a *Salmonella* epidemic (Felix 1956), immobilizing the phage to polystyrene in a dipstick or microtiter plate format by soaking the surfaces with phage suspensions, then washing to remove unbound phage and remaining absorption sites. The immobilized phage were incubated with *Salmonella* and a mixed bacterial culture, and subsequent analysis by PCR suggested the specific recovery of nine out of eleven *Salmonella* samples (Bennett et al. 1997). However, a concentration of 105 colony forming units (cfu) per milliliter was required to obtain a positive result by PCR. Overall, this accounted for a capture efficiency of only one percent of cells, limiting the tests sensitivity and clinical utility (Bennett et al. 1997; Schmelcher and Loessner 2014).

Chemical functionalization of bacteriophages to surfaces can dramatically enhance the passive absorption method of phage described above. Phage functionalization using sugars such as dextrose and sucrose, or amino acids such as histidine and cysteine, has been found to greatly improve the number of phage bound to a variety of surfaces (Singh et al. 2012). Handa et al. (2008) reported a method to chemically immobilize phage P22 in a monolayer with application toward the detection of various bacteria including *S. typhimurium*, *E. coli*, and *Listeria monocytogenes* (Handa et al. 2008). The immobilized phage showed particularly strong binding to *S. typhimurium*, offering promise for this method to be applied as a biosensor technique.

Phages have also been modified to display ligands on their heads for an oriented immobilization onto surfaces. This is important as the tail of the phage must interact with the bacterium in order to adsorb to its host cell. This oriented immobilization has been shown in a study by Gervais et al. (2007) where T4 phages were immobilized onto gold surfaces through genetic biotinylation of the phage heads. This process involved displaying biotin on the surface of the phage heads to interact with the streptavidin-coated magnetic beads for an oriented immobilization. Here, the phage were demonstrated to be able to capture up to 99 % of their targeted hosts from suspension (Gervais et al. 2007).

5 Conclusions

Bacteriophage-based detection techniques offer great potential to circumvent the biological threats imparted by bacterial pathogens. Plaque forming phages have been used extensively for bacterial detection, particularly in combination with phage display techniques, to aid in the targeting of specific pathogens. Reporter phage assays have further sophisticated phage reporter strategies thus, improving the speed and sensitivity of detection. Further enhancements involving the oriented immobilization of the reporter phage will aid to push the limits of detection and sensitivity.

References

Barry, M., Dower, W., & Johnston, S. A. (1996). Toward cell-targetting gene therapy vectors: Selection of cell-binding peptides from random peptide-presenting phage libraries. *Nature Medicine, 2*(3), 299–305.

Bennett, A. R., Davids, F. G., Vlahodimou, S., Banks, J. G., & Betts, R. P. (1997). The use of bacteriophage-based systems for the separation and concentration of Salmonella. *Journal of Applied Microbiology, 83*(2), 259–265. Retrieved from http://www.ncbi.nlm.nih.gov/pubmed/9281830

Berkane, E., Orlik, F., Stegmeier, J. F., Charbit, A., Winterhalter, M., & Benz, R. (2006). Interaction of bacteriophage lambda with its cell surface receptor: An in vitro study of binding of the viral tail protein gpJ to LamB (Maltoporin). *Biochemistry, 45*(8), 2708–2720. http://doi.org/10.1021/bi051800v

Brigati, J., Williams, D. D., Sorokulova, I. B., Nanduri, V., Chen, I.-H., Turnbough, C. L., et al. (2004). Diagnostic probes for *Bacillus anthracis* spores selected from a landscape phage library. *Clinical Chemistry, 50*(10), 1899–1906.

Brokopp, C. D., Gomez-Lus, R., & Farmer, J. J. (1977). Serological typing of *Pseudomonas aeruginosa*: Use of commercial antisera and live antigens. *Journal of Clinical Microbiology, 5*(6), 640–649.

Brovko, L. Y., Anany, H., & Griffiths, M. W. (2012). Bacteriophages for detection and control of bacterial pathogens in food and food-processing environment. *Advances in Food and Nutrition Research, 67*, 241–288. http://doi.org/10.1016/B978-0-12-394598-3.00006-X

Broxmeyer, L., Sosnowska, D., Miltner, E., Chacón, O., Wagner, D., McGarvey, J., ... Bermudez, L. E. (2002). Killing of *Mycobacterium avium* and *Mycobacterium tuberculosis* by a mycobacteriophage delivered by a nonvirulent mycobacterium: a model for phage therapy of intracellular bacterial pathogens. *The Journal of Infectious Diseases, 186*(8), 1155–1160.

Cox, C. (2012). Bacteriophage-based methods of bacterial detection and identification. In P. Hyman & S. T. Abedon (Eds.), *Bacteriophages in Health and Disease* (pp. 134–152). Advances in Molecular and Cellular Microbiology.

Felix, A. (1956). Phage typing of *Salmonella typhimurium*: Its place in epidemiological and epizootiological investigations. *Journal of General Microbiology, 14*, 208–222.

Funatsu, T., Taniyama, T., Tajima, T., Tadakuma, H., & Namiki, H. (2002). Rapid and sensitive detection method of a bacterium by using a GFP reporter phage. *Microbiology and Immunology, 46*(6), 365–369.

Gervais, L., Gel, M., Allain, B., Tolba, M., Brovko, L., Zourob, M., ... Evoy, S. (2007). Immobilization of biotinylated bacteriophages on biosensor surfaces. *Sensors and Actuators B: Chemical, 125*(2), 615–621.

Gillis, A., & Mahillon, J. (2014). Phages preying on *Bacillus anthracis*, *Bacillus cereus*, and *Bacillus thuringiensis*: Past, present and future. *Viruses, 6*(7), 2623–2672.

Gopinath, S. C. B., Tang, T.-H., Chen, Y., Citartan, M., & Lakshmipriya, T. (2014). Bacterial detection: From microscope to smartphone. *Biosensors and Bioelectronics, 60*, 332–342.

Grajewski, B. A., Kusek, J. W., & Gelfand, H. M. (1985). Development of a bacteriophage typing system for *Campylobacter jejuni* and *Campylobacter coli*. *Journal of Clinical Microbiology, 22*(1), 13–18.

Gulig, P., Martin, J., Messer, H. G., Deffense, B. L., & Harpley, C. J. (2008). Phage display methods for detection of bacterial pathogens. In M. Zourpb, S. Elwary, & A. Turner (Eds.), *Principles of bacterial detection: Biosensors, recognition receptors and microsystems* (pp. 755–783). Montreal, Canada: Springer.

Handa, H., Gurczynski, S., Jackson, M. P., Auner, G., & Mao, G. (2008). Recognition of *Salmonella typhimurium* by immobilized phage P22 monolayers. *Surface Science, 602*(7), 1392–1400.

Housby, J. N., & Mann, N. H. (2009). Phage therapy. *Drug Discovery Today, 14*(11–12), 536–540.

Ide, T., Bik, S., Matsuba, T., & Marayama, S. H. (2003). Identification by the phage-display technique of peptides that bind to H7 flagellin of *Escherichia coli*. *Bioscience, Biotechnology, and Biochemistry, 67*(6), 1335–1341.

Kalniņa, Z., Siliņa, K., Meistere, I., Zayakin, P., Rivosh, A., Abols, A., ... Linē, A. (2008). Evaluation of T7 and lambda phage display systems for survey of autoantibody profiles in cancer patients. *Journal of Immunological Methods, 334*(1–2), 37–50,

Loessner, M. J., Rees, C. E., Stewart, G. S., & Scherer, S. (1996). Construction of luciferase reporter bacteriophage A511:luxAB for rapid and sensitive detection of viable *Listeria* cells. *Applied and Environmental Microbiology, 62*(4), 1133–1140.

Loessner, M. J., Rudolf, M., Scherer, S., & Icrobiol, A. P. P. L. E. N. M. (1997). Evaluation of luciferase reporter bacteriophage A511: luxAB for detection of *Listeria monocytogenes* in contaminated foods †. *Microbiology, 63*(8), 2961–2965.

Minion, J., Leung, E., Menzies, D., & Pai, M. (2010). Microscopic-observation drug susceptibility and thin layer agar assays for the detection of drug resistant tuberculosis: A systematic review and meta-analysis. *The Lancet Infectious Diseases, 10*(10), 688–698. http://doi.org/10.1016/S1473-3099(10)70165-1

Nicastro, J., Sheldon, K., El-Zarkout, F. A., Sokolenko, S., Aucoin, M. G., & Slavcev, R. (2013). Construction and analysis of a genetically tuneable lytic phage display system. *Applied Microbiology and Biotechnology, 97*(17), 7791–7804.

Oda, M., Morita, M., Unno, H., & Tanji, Y. (2004). Rapid detection of *Escherichia coli* O157: H7 by using green fluorescent protein-labeled PP01 bacteriophage. *Applied and Environmental Microbiology, 70*(1), 527–534.

Olsen, E. V., Sorokulova, I. B., Petrenko, V. A., Chen, I.-H., Barbaree, J. M., & Vodyanoy, V. J. (2006). Affinity-selected filamentous bacteriophage as a probe for acoustic wave biodetectors of *Salmonella typhimurium*. *Biosensors and Bioelectronics, 21*(8), 1434–1442.

Ripp, S., Jegier, P., Birmele, M., Johnson, C. M., Daumer, K. A., Garland, J. L., et al. (2006). Linking bacteriophage infection to quorum sensing signalling and bioluminescent bioreporter monitoring for direct detection of bacterial agents. *Journal of Applied Microbiology, 100*, 488–499.

Rishi, P., Singh, A. P., Arora, S., Garg, N., & Kaur, I. P. (2014). Revisiting eukaryotic anti-infective biotherapeutics. *Critical Reviews in Microbiology, 40*(4), 281–292.

Schmelcher, M., & Loessner, M. J. (2014). Application of bacteriophages for detection of foodborne pathogens. *Bacteriophage, 4*(2), e28137. http://doi.org/10.4161/bact.28137

Singh, A., Arutyunov, D., Szymanski, C. M., & Evoy, S. (2012). Bacteriophage based probes for pathogen detection. *The Analyst, 137*, 3405.

Singh, P., Gonzalez, M. J., & Manchester, M. (2006). Viruses and their uses in nanotechnology. *Nanotechnology, 67*, 23–41. http://doi.org/10.1002/ddr

Smith, G. P., & Petrenko, V. A. (1997). Phage display. *Chemical Reviews, 97*(2), 391–410.

Sorokulova, I. B., Olsen, E. V, Chen, I.-H., Fiebor, B., Barbaree, J. M., Vodyanoy, V. J., ... Petrenko, V. A. (2005). Landscape phage probes for *Salmonella typhimurium*. *Journal of Microbiological Methods, 63*(1), 55–72.

Stewart, G., Jassim, S., Denyer, S. P., Newby, P., Linley, K., & Dhir, V. (1998). The specific and 329 sensitive detection of bacterial pathogens within 4 h using bacteriophage amplification. *Journal of Applied Microbiology, 84*(5), 777–783.

Tanji, Y., Shimada, T., Yoichi, M., Miyanaga, K., Hori, K., & Unno, H. (2004). Toward rational control of *Escherichia coli* O157:H7 by a phage cocktail. *Applied Microbiology and Biotechnology, 64*(2), 270–274.

Tawil, N., Sacher, E., Mandeville, R., & Meunier, M. (2014). Bacteriophages: Biosensing tools for multi-drug resistant pathogens. *The Analyst, 139*(6), 1224–1236. http://doi.org/10.1039/c3an01989f

Thal, E., & Nordberg, B. K. (1968). On the diagnostic of *Bacillus anthracis* with bacteriophages. *Berliner Und Münchener Tierärztliche Wochenschrift, 81*(1), 11–13.

Turnbough, C. L. (2003). Discovery of phage display peptide ligands for species-specific detection of *Bacillus* spores. *Journal of Microbiological Methods, 53*(2), 263–271.

Ulitzur, N., & Ulitzur, S. (2006). New rapid and simple methods for detection of bacteria and determination of their antibiotic susceptibility by using phage mutants new rapid and simple methods for detection of bacteria and determination of their antibiotic susceptibility by using phage M. *Society.*

Ulitzur, S., & Kuhn, J. (1987). Introduction of lux genes into bacteria, a new approach for specific determination of bacteria and their antibiotic susceptibility. In J. Chlomerich, R. Andreesen, A. Kapp, M. Ernst, & W. Wood (Eds.), *Bioluminescence and chemiluminescence, new perspectives.* (pp. 463–72). Wiley.

Vesuna, F., Winnard, P., & Raman, V. (2005). Enhanced green fluorescent protein as an alternative control reporter to *Renilla luciferase. Analytical Biochemistry, 342*(2), 345–347.

Wallmark, G., Arremark, I., & Telander, B. (1978). *Staphylococcus saprophyticus*: A frequent cause of acute urinary tract infection among female outpatients. *The Journal of Infectious Diseases, 138*(6), 791–797.

Chapter 7
Phage-Mediated Immunomodulation

1 Introduction

While the natural hosts for bacteriophages are bacteria, there is growing evidence for the ability of phage to interact with mammalian cells, particularly with those of the human immune system. These interactions typically encompass two main features: (i) phage immunogenicity, or ability of phages to induce specific immune responses; and (ii) phage immunomodulation, which can be defined as the ability of phages to modify the immune system in both innate and adaptive responses. The aim of this chapter is to explore the interactions between phages and the immune system, and more specifically the implications of these interactions in the development of novel medical applications.

2 Immune Responses to Phage

The pervasive presence of bacteriophages in the environment is indicative of the constant exposure of humans to phages. Not surprisingly, phages are well tolerated by mammalian hosts. This tolerance comes with the stimulation of various immune responses from that mammal that are critical to consider in the development of any therapeutic. To the eukaryotic host, phages are highly immunogenic foreign entities that interact with the innate immune system and induce specific humoral and cellular immune responses (Górski et al. 2005; Kaur et al. 2012). Despite the intriguing evidence for bacteriophage-initiated immune responses, there is a limited understanding of how phage exposure triggers these responses or how they can ultimately be expended to influence immunity in mammalian systems.

© The Author(s) 2016
J. Nicastro et al., *Bacteriophage Applications—Historical Perspective and Future Potential*, SpringerBriefs in Biochemistry and Molecular Biology, DOI 10.1007/978-3-319-45791-8_7

Some plausible theories have been developed describing the routes for phage penetration, based on route of administration and immune activation. While phage penetration studies have been inconsistent, the evidence for phage entry into the bloodstream and systemic tissues is consistent in nearly every animal model and administration route. Administration routes considered to date include: oral (Delmastro et al. 1997; March et al. 2004; Zuercher et al. 2004), subcutaneous (Grabowska et al. 2000; Wu et al. 2002), intramuscular (Acton and Evans 1968; Clark et al. 2011; Clark and March 2004; Samoylov et al. 2015), intravenous (Inchley 1969), and intradermal injection (Roehnisch et al. 2013). Animal models studied to date include: mice (Samoylov et al. 2015; Willis et al. 1993; Wu et al. 2002), rabbits (Clark et al. 2011; March et al. 2004), pigs (Zuercher et al. 2004), non-human primates (Chen et al. 2001; Rao et al. 2004; Trouche et al. 2009), as well human in phase I/II clinical trials for the treatment of patients with multiple myeloma (Roehnisch et al. 2013, 2014). This section will consider the specific immune responses that inactivate phage, specifically focusing on innate responses clearing phage and the generation of specific antibodies against phage antigens.

2.1 Anti-phage Innate Responses

The innate immune response can be defined as an immediate, fast-acting, first-line antimicrobial host defense that remains active until a more definitive adaptive response can be developed. Both systems are interactive and therefore cannot be considered as completely separate entities (Fearon and Locksley 1996). As foreign, typically exogenous entities, bacteriophages do initiate innate immune responses with many similarities to their bacterial counterparts. This section will discuss immune responses to phage, specifically involving phage clearance and neutralization.

Bacteriophages are highly immunogenic particles that are recognized as foreign entities to eukaryotic hosts and as such, are quickly removed from the body when no bacterial hosts remain to be infected (Olszowska-Zaremba et al. 2012a, b). The system most responsible for phage clearance is the reticuloendothelial system (RES) of the liver and spleen (Dabrowska et al. 2005; Geier et al. 1973; Inchley 1969). It is a component of the innate immune system comprised of phagocytic cells located in the reticular connective tissue (Halpern 1959). These cells are primarily monocytes and macrophages, though Kupffer cells (mononuclear phagocytes of the liver) and dendritic cells have also been suggested to play a significant role in the rapid removal of phage. The responses by Kupffer cells to T4 phage were first studied by Inchley (1969), where the investigator noted that T4 was rapidly removed from the livers and spleens of mice after being phagocytosed by Kupffer cells. The phagocytosis was noted as early as 30 minutes post-infection and while macrophages of the spleen were also tested for their ability to remove phage, the ability of the RES immune cells to clear the phage was comparatively much lower than that of Kupffer cells.

A more recent study by Molenaar et al. (2002) has similarly suggested a specificity of phages to Kupffer cells, where the pharmacokinetics and processing of filamentous M13 phage was investigated in mice. Internalization of the M13 phage via receptor-mediated endocytosis was observed within 30 minutes of infection, and the immune responses of the RES were further exemplified by a 5000-fold reduction of bioactive phage within 90 minutes of infection (Molenaar et al. 2002).

The rapid mammalian clearance of bacteriophages can also be associated with the physico-chemical properties of their surface proteins, which are likely responsible for their recognition as foreign particles. Merril et al. (1996) attempted to circumvent the innate responses that were involved with the rapid removal of phage P22 and phage λ in mice models to create "long-circulating" phage mutants. The elegant system involved the selection of mutants, or residual phage, through the continued isolation of phage that could progressively evade the RES as demonstrated by their ability to passage through and persist in the mice.

2.2   Humoral Immune Reposes to Phage (Anti-phage Antibodies)

Anti-phage-neutralizing antibodies are undeniably one of the most important limiting factors in phage-based therapies. Exogenous phage have been shown to induce both innate and antibody responses in their mammalian hosts—a response that is partially impacted by the type of phage administered, the route of administration and dosage protocol (Sulakvelidze et al. 2001). Topical and oral administrations will typically result in lower antibody responses in comparison to other methods (Olszowska-Zaremba et al. 2012a, b).

Neutralizing antibodies can greatly decrease the efficacy of bacteriophage-based therapies, making the identification of immunogenic phage components of particular interest. Antibody generation against filamentous bacteriophage was investigated in a study by van Houten et al. (2010) after they had identified filamentous phage as highly immunogenic agents (van Houten et al. 2006). The group found that deletion mutants of the N1 and N2 domains of the pIII coat protein were significantly less immunogenic in comparison to wild-type M13 filamentous phage as well as those with altered pVIII coat proteins (van Houten et al. 2010).

In another M13 phage immunization study, Frenkel et al. (2000) demonstrated evidence for the development of serum IgG antibodies against recombinant M13 phage (Frenkel et al. 2000). This work was further studied and confirmed by Hashiguchi et al. (2010), where primary antibody responses were found to consist of both IgM and IgG isotypes (Hashiguchi et al. 2010). The single stranded nature of the M13 DNA genome possesses unmethylated dinucleotide CpG motifs that may also impart immunological effects (Karimi et al. 2016).

Evidence for antibody generation to T7 phage was demonstrated by Sokoloff et al. (2000), where administered T7 phages were recognized by natural antibodies in the circulatory systems of their rat models and were effectively removed following this recognition. Interestingly, the group noted that rate of phage survival was specific to the peptide being displayed on the phage, suggesting that lysine and arginine residues can protect the phage against humoral immune responses (Sokoloff et al. 2000). Further studies by Srivastava et al. (2004) showed that the clearance of T7 phage from the blood is slower in B-cell deficient mice as compared to wild-type mice and that this clearance can vary depending on the type of phage (Srivastava et al. 2004; Sulakvelidze 2005).

2.3   Anti-phage Cellular Immunity and the Implications of the Impact of Phage on the Adaptive Responses (T and B Cells)

Cellular immunity plays an essential role in combating viral infections. In particular, T-cells are required to expand virus-specific memory-dependent cytotoxic and helper, CD8+ and CD4+ T-cells (Watanabe et al. 1992). Phages that have undergone endocytosis by specialized antigen-presenting cells (APCs) described above, would be presented to antigen-specific CD4+ T-helper cells, imparting the generation of memory cells against the antigenic determinants of a phage, resulting in the rapid clearance upon subsequent exposure through neutralizing antibodies (Kaur et al. 2012). This phenomenon has been demonstrated through the use of phage ΦX174, which has been used extensively in immunodeficiency screenings (Kaur et al. 2012; Lopez et al. 1975; Ochs et al. 1971). Ulivieri et al. (2008) described the processing of recombinant M13 bacteriophage by human antigen-presenting cells onto MHC class II molecules. In an earlier study on the topic by Yang et al. (2005), the authors suggested a link between the demonstrated antibody responses and the activation of the Th1 cytotoxic T-cells and Th2 T-helper cell responses.

3   Bacteriophage—Based Immune-Pharmaco-Therapies

The natural immunostimulatory effects of bacteriophage, in combination with their versatility, offers the capacity for an assortment of exciting immunological therapies. While the applications of phage toward immune-pharmaco-therapies is virtually limitless, here we will focus on applications to cancer with other considerations for autoimmune disorders, drug addiction, obesity and reactive oxygen species derived as a result of bacterial infections.

3.1 Phage Immunogenicity and Cancer Therapy

Cancer is a highly complex and multifactorial disease. It is generally defined by several definitive hallmarks (Hanahan and Weinberg 2000, 2011) including, but not limited to: rapid over-proliferation of cells, enhanced proliferative signaling, and angiogenesis. Cancer therapies therefore generally seek to eliminate such behaviors (killing rapidly proliferating cells, for example) or target them to limit exposure of normal tissues to the lethal cancer therapeutic. The ideal cancer therapy seeks to selectively target and eliminate cancerous cells while preserving normal, non-cancer cells (Strebhardt and Ullrich 2008).

3.1.1 Phage in Tumor Targeting

Targeted therapy has historically made use of monoclonal antibodies that confer high affinity for tumor antigens. However, their large size (~ 160 kDa) (Jain 1990) and their high affinity for the initial antigens with which they ligate (Adams et al. 2001) may actually reduce the extent to which they are able to penetrate a tumor, thereby reducing efficacy. As such, smaller antibody and peptide fragments may be more desirable as tumor targeting moieties. Phage display libraries (see Smith and Petrenko (1997) for a review of phage display) are indispensable for high-throughput and cost-effective selection of potential peptides amongst billions of candidates (Hamzeh-Mivehroud et al. 2013; Ladner et al. 2004). High affinity ligands isolated from phage display libraries have been demonstrated as probes for cancer detection (Newton et al. 2006; Samoylova et al. 2003) and in the improvement of targeted delivery (liposomal) (Bedi et al. 2014; Jayanna et al. 2009, 2010a, b; Wang et al. 2011) of therapeutic cargo. Identification of cell penetrating peptides (CPPs) (Barry et al. 1996), tumor-specific CPP in particular (Ivanenkov et al. 1999; Poul et al. 2000), has also been successful, which is particularly important for oligonucleotide or organelle-specific drug delivery.

Phage display has also been used to screen for potential cell surface receptors as novel markers to be exploited in cancer therapy (Molek et al. 2011). The interleukin-11 receptor α was first identified as a promising target for metastatic prostate cancer through an in vivo combinatorial screening approach (Cardó-Vila et al. 2008; Lewis et al. 2009; Zurita 2004). Similarly, a ligand-directed peptide drug has since been developed against the receptor with very promising results in the recent first-in-man study (Pasqualini et al. 2015).

Tumor heterogeneity is a recognized obstacle in the development of targeted cancer therapies. Gross et al. (2016) recently described the use of a landscape phage display library to identify "promiscuous" ligands, which are defined as targeting ligands with affinity for multiple receptors across multiple cancer phenotypes. As such, this group identified phage ligands against heterogeneous pancreatic and lung cancer cell populations. Phage particles displaying the ligand were able to rapidly accumulate within target cells, which, they proposed, was due to the interaction of the "promiscuous" ligands with multiple cell surface receptors. Phage display will likely continue to lead the way in the identification of future similar peptides for more complex arrays of targets.

3.1.2 Anti-tumor Phage Therapies

Phage display may be established as a fundamental tool in the discovery of tumor targeting ligands, but phage particles themselves are also being investigated as therapeutics. Due to their small size and simple genomics, phage offer the capacity to serve as excellent candidates for novel nanomedicines (Kaur et al. 2012; Moghimi et al. 2005). For example, it has been demonstrated that surface residues on fila-mentous phage M13 can be functionalized for bioconjugation to other moieties (Henry et al. 2015; Niu et al. 2008; see Petrenko and Jayanna (2014) and Kaur et al. (2012) for more extensive reviews on existing phage-based nanomedicines outside of cancer).

The greatest potential for phage cancer therapy may lie in their capacity to carry therapeutic cargo (DePorter and McNaughton 2014; Henry et al. 2015), which takes advantage of phage tumor-targeting ligand display. Furthermore, phage particles remain stable across both a wide temperature and pH range, which is desirable for a drug delivery vehicle (Jonczyk et al. 2011). Samoylova et al. (2003) constructed phage particles expressing three families of glioma-specific ligands. These phage were highly specific against a malignant glioma cell line and the authors suggested their use as probes to identify cell surface markers in glioma cases. Phage that display targeting ligands can be taken up by mammalian cells through endocytosis in vitro and in vivo (Kassner et al. 1999; Larocca et al. 2001; Poul and Marks 1999; Urbanelli et al. 2001), which is desirable for intracellular delivery of oligonu-cleotides or organelle-targeted small molecules. CPPs selected on phage display libraries can be utilized to enhance cell-specific internalization of phage, as in the "internalizing phage" (iPhage) system by Rangel et al. (2013). The CPP penetratin (pen), derived from *Drosophila* antennapedia (Derossi et al. 1994, 1996), is fused to the major coat protein pVIII of filamentous phage M13 (Rangel et al. 2012), enabling its receptor-independent entry into mammalian cells. The iPhage system then enables fusion of a targeting peptide to the minor coat protein pIII for ligand-directed intracellular targeting.

One promising approach to phage-mediated gene delivery is through phagemid infective particles (PIPs) (Mount et al. 2004). This system builds off of the com-position of a minimal phagemid encoding only the minimal genetic elements necessary for bacterial and mammalian cell replication as well as the therapeutic gene to be delivered. All other phage packaging and assembly proteins are encoded by a helper phage present in the same host bacterial cell. Most relevantly, this includes a coat protein fused with a targeting ligand, previously selected from a cell-targeted phage library. The resulting phage particles (PIPs) produced from the bacterial host are therefore usually smaller than the typical phage. These PIPs were demonstrated to bind to and be taken up by target cells where they then release their genetic cargo for processing. By separating the genes for encapsulation and targeting from the infecting phagemid, Mount et al. (2004) developed a modular system with great flexibility with respect to targeting and therapeutic cargo, where PIPs encoding GFP were convincingly targeted and delivered to metastatic prostate cancer cells (Fagbohun et al. 2013).

First observed by Bloch (1940), phage can preferentially accumulate within solid tumors and inhibit tumor growth (Dabrowska et al. 2004a, b, 2005). A putative mechanism was proposed based on the interaction between phage capsid proteins and the integrin receptors on the cell surface (Dabrowska et al. 2004a, b). The intrinsic immunogenicity of accumulated phage within tumor environments could hypothetically induce a strong local inflammatory response to mediate tumor destruction. However, it is also likely that the natural immunogenicity of phages may also elicit the formation of neutralizing antibodies, thereby reducing their effect upon repeated administration (Kaur et al. 2012). Specific immunostimulation against cancer cells is, however, a highly promising approach (Rosenberg 1999) to eradicate tumor cells by exploiting the presence of cancer-specific antigens. The use of phage immunomodulation has also already been discussed in the development of vaccines. A similar approach can be taken against tumor antigens to elicit a protective anti-tumor immune response. Wu et al. (2002) demonstrated the potential of filamentous phage displaying tumor antigens (P1A35-43) to prime an immune response in a murine mastocytoma tumor model. Administration of the phage particles elicited a P1A35-43 specific CD8+ T lymphocyte response and induced IFN-γ production. Similarly, Fang et al. (2005) administered phage displaying melanoma-specific antigens (MAGE-A1) to mice and also observed a protective anti-tumor effect, specifically elevated CTL response specific to MAGE-A1 and elevated NK activity. Eriksson et al. (2007, 2009) also engineered a tumor-specific filamentous phage through in vivo phage display that specifically localized within tumors. They demonstrated accumulation of the phage within tumor environments in a melanoma murine model. Immunogenic phage induced high levels of cytokines, including IL-12 and IFN-γ. They observed tumor reduction, delayed tumor progression, and prolonged lifespan of tumor-bearing mice.

3.2 Bacteriophage Immunotherapy Autoimmune Disorders

The development of the phage display technique has provided the opportunity to clone and characterize human immune libraries. This has greatly facilitated our understanding of autoimmune diseases (Bazan et al. 2012), including developing evidence for antigen-autoantibody reactions with TTP (Luken et al. 2006) and AAU (Kim et al. 2011) and other autoimmune ocular inflammatory disorders (Bazan et al. 2012). Phage display has proven useful in designing therapeutic agents against autoimmune diseases, including the generation of galectin-3 mimotopes that could serve to regulate immune responses in Crohn's patients with increased anti-galectin-3 IgG autoantibodies in their sera, compared to healthy patients (Bazan et al. 2012; Sblattero et al. 1999).

Another exciting application for phage-based autoimmune therapies was conducted by Kolonin et al. (2004) who used phage display in vivo as a potential therapeutic for anti-obesity therapy. Here, bacteriophages were developed to target proapoptotic CKGGRAKDC peptide to prohibit the adipose vasculature, which resulted in the ablation of white fat and a reported normalization of metabolism with no apparent adverse effects (Bazan et al. 2012; Kolonin et al. 2004).

3.3   Bacteriophage Immunotherapy for Drug Addiction

Drug addiction is an important health and social problem across the world that imparts massive economic and social consequences. Among the habitually abused drugs, cocaine may be the most addictive and as such represents an investigative target for new treatments and therapies. It has been demonstrated that, protein-based therapeutics, in which proteins are designed to bind to cocaine, could reduce the load of cocaine and consequently eliminate its psychoactive effects. To date this strategy has not proven significant due to the inability of these proteins to directly access the CNS by crossing the blood-brain barrier. To address this limitation, Carrera et al. (2004) engineered filamentous bacteriophage, a phage that has shown the ability to pass through the blood-brain barrier and to access the central nervous system (CNS) to display cocaine sequestrating antibodies on its surface, thereby neutralizing the drug in the brain. The modified phage were administered intra-nasally to rats twice a day for three consecutive days at a high titer of 1×10^{15} phage. Brain samples were harvested throughout treatment and after to assess phage penetrance and accumulation. Overall, titers greater than 10^9 were detected within seven days of the treatment with recombinant phage, peaking at 2.5×10^{13} on day 4. This confirms the ability of filamentous bacteriophage to penetrate the blood-brain barrier and enter the CNS. The results of this study support the development of a new system for treatment of cocaine addiction and also demonstrate that this strategy could serve as a platform for drug abuse treatment (Clark and March 2006; Dickerson et al. 2005).

3.4   Phages and Oxidative Stress

An important factor to consider in non-specific immune responses to phage involves the phagocytosis of microbes, in which Reactive Oxygen Species (ROS) are generated to remove bacterial threats. ROS species involved in this response include: hydrogen peroxide, superoxide, and hydroxyl radicals. When ROS are produced in excess they can induce oxidative stress and cytotoxic activity and are implicated in the pathogenesis of cancer, sepsis, multiple organ failure and neurodegenerative disorders (Bartsch and Nair 2006; Olszowska-Zaremba et al. 2012a, b). As such, the ROS-suppressive effects found in bacteriophage are desirable and warrant further investigation. Pzerwa et al. (2006) demonstrated the ability of bacteriophages to inhibit the degree of chemiluminescence and hence, the rate of phagocytosis in mouse phagocytes even when incubated with the previous inducers of the immune response, such as the lipopolysaccharide (LPS) of *E. coli*.

4 Conclusions

Similar to their mammalian viral counterparts, bacteriophages are nucleo-proteinaceous in composition and as such, are recognized by mammalian hosts as foreign particles and are subject to immunostimulatory effects that aim to clear them (Kaur et al. 2012; Rishi et al. 2014; Srivastava et al. 2004). The use of the safe bacteriophages in new immunotherapies is a natural progression and offers great potential to the field of nanomedicine. However, the use of bacteriophages in mammals must also undergo careful consideration in terms of developing a more thorough understanding of mammalian host-specific immune responses as well as the phage pharmacokinetics (PK) and pharmacodynamics involved with bacterio-phage therapies (Abedon and Thomas-Abedon 2010).

References

Abedon, S. T., & Thomas-Abedon, C. (2010). Phage therapy pharmacology. *Current Pharmaceutical Biotechnology, 11*(1), 28–47.

Acton, R. T., & Evans, E. E. (1968). Bacteriophage clearance in the (*Crassostrea virginica*). *Journal of Bacteriology, 95*(4), 1260–1266.

Adams, G. P., Schier, R., McCall, A. M., Simmons, H. H., Horak, E. M., Alpaugh, R. K., … Weiner, L. M. (2001). High affinity restricts the localization and tumor penetration of single-chain Fv antibody molecules. *Cancer Research, 61*(12), 4750–4755.

Barry, M. A., Dower, W. J., & Johnston, S. A. (1996). Toward cell–targeting gene therapy vectors: Selection of cell–binding peptides from random peptide–presenting phage libraries. *Nature Medicine, 2*(3), 299–305.

Bartsch, H., & Nair, J. (2006). Chronic inflammation and oxidative stress in the genesis and perpetuation of cancer: Role of lipid peroxidation, DNA damage, and repair. *Langenbeck's Archives of Surgery/ Deutsche Gesellschaft Für Chirurgie, 391*(5), 499–510.

Bazan, J., Całkosiński, I., & Gamian, A. (2012). Phage display—A powerful technique for immunotherapy. *Human Vaccines and Immunotherapeutics, 8*, 1829–1835.

Bedi, D., Gillespie, J. W., & Petrenko, V. A. (2014). Selection of pancreatic cancer cell-binding landscape phages and their use in development of anticancer nanomedicines. *Protein Engineering, Design and Selection, 27*(7), 235–243.

Bloch, H. (1940). Experimental investigation of the relationship between bacteriophage and malignant tumors. *Arch Gesamte Virusforsch, 1*, 481–496.

Cardó-Vila, M., Zurita, A. J., Giordano, R. J., Sun, J., Rangel, R., Guzman-Rojas, L., … Pasqualini, R. (2008). A ligand peptide motif selected from a cancer patient is a receptor-interacting site within human interleukin-11. *PloS One, 3*(10), e3452.

Carrera, M. R. A., Kaufmann, G. F., Mee, J. M., Meijler, M. M., Koob, G. F., Janda, K. D. (2004). Treating cocaine addiction with viruses. *Proceedings of the National Academy of Sciences of the United States of America, 101*(28), 10416–10421. doi: 10.1073/pnas.0403795101

Chen, X., Scala, G., Quinto, I., Liu, W., Chun, T. W., Justement, J. S., … Fauci, a S. (2001). Protection of rhesus macaques against disease progression from pathogenic SHIV-89.6PD by vaccination with phage-displayed HIV-1 epitopes. *Nature Medicine, 7*(11), 1225–1231.

Clark, J. R., Bartley, K., Jepson, C. D., Craik, V., & March, J. B. (2011). Comparison of a bacteriophage-delivered DNA vaccine and a commercially available recombinant protein vaccine against hepatitis B. *FEMS Immunology and Medical Microbiology, 61*(2), 197–204.

Clark, J. R., & March, J. B. (2004). Bacteriophage-mediated nucleic acid immunisation. *FEMS Immunology and Medical Microbiology, 40*, 21–26.

Clark, J. R., & March, J. B. (2006). Bacteriophages and biotechnology: Vaccines, gene therapy and antibacterials. *Trends in Biotechnology, 24*(5), 212–218.

Dabrowska, K., Opolski, A., Wietrzyk, J., Switala-Jelen, K., Boratynski, J., Nasulewicz, A., … Gorski, A. (2004a). Antitumor activity of bacteriophages in murine experimental cancer models caused possibly by inhibition of beta3 integrin signaling pathway. *Acta Virologica, 48*(4), 241–248.

Dabrowska, K., Opolski, A., Wietrzyk, J., Switala-Jelen, K., Godlewska, J., Boratynski, J., … Gorski, A. (2004b). Anticancer activity of bacteriophage T4 and its mutant HAP1 in mouse experimental tumor models. *Anticancer Research, 24*(6), 3991–3995.

Dabrowska, K., Switala-Jelen, K., Opolski, A., Weber-Dabrowska, B., & Gorski, A. (2005). Bacteriophage penetration in vertebrates. *Journal of Applied Microbiology, 98*(1), 7–13.

Delmastro, P., Meola, A., Monaci, P., Cortese, R., & Galfrè, G. (1997). Immunogenicity of filamentous phage displaying peptide mimotopes after oral administration. *Vaccine, 15*(11), 1276–1285.

DePorter, S. M., & McNaughton, B. R. (2014). Engineered M13 bacteriophage nanocarriers for intracellular delivery of exogenous proteins to human prostate cancer cells. *Bioconjugate Chemistry, 25*(9), 1620–1625.

Derossi, D., Joliot, A. H., Chassaing, G., & Prochiantz, A. (1994). The third helix of the antennapedia homeodomain translocates through biological membranes. *Journal of Biological Chemistry, 269*(14), 10444–10450.

Derossi, D., Calvet, S., Trembleau, A., Brunissen, A., Chassaing, G., & Prochiantz, A. (1996). Cell internalization of the third helix of the antennapedia homeodomain is receptor-independent. *Journal of Biological Chemistry, 271*(30), 18188–18193.

Dickerson, T. J., Kaufmann, G. F., & Janda, K. D. (2005). Bacteriophage-mediated protein delivery into the central nervous system and its application in immunopharmacotherapy. *Expert Opinion on Biological Therapy, 5*(6), 773–781.

Eriksson, F., Culp, W. D., Massey, R., Egevad, L., Garland, D., Persson, M. A. A., et al. (2007). Tumor specific phage particles promote tumor regression in a mouse melanoma model. *Cancer Immunology, Immunotherapy, 56*(5), 677–687.

Eriksson, F., Tsagozis, P., Lundberg, K., Parsa, R., Mangsbo, S. M., Persson, M. A. A., … Pisa, P. (2009). Tumor-specific bacteriophages induce tumor destruction through activation of tumor-associated macrophages. *Journal of Immunology, 182*(5), 3105–3111.

Fagbohun, O. A, Kazmierczak, R. A, Petrenko, V. A, & Eisenstark, A. (2013). Metastatic prostate cancer cell-specific phage-like particles as a targeted gene-delivery system. *Journal of Nanobiotechnology, 11*(1), 31. http://doi.org/10.1186/1477-3155-11-31

Fang, J., Wang, G., Yang, Q., Song, J., Wang, Y., & Wang, L. (2005). The potential of phage display virions expressing malignant tumor specific antigen MAGE-A1 epitope in murine model. *Vaccine, 23*(40), 4860–4866.

Fearon, D. T., & Locksley, R. M. (1996). The instructive role of innate immunity in the acquired immune response. *Science, 272*(5258), 50–54.

Frenkel, D., Katz, O., & Solomon, B. (2000). Immunization against Alzheimer's beta-amyloid plaques via EFRH phage administration. *Proceedings of the National Academy of Sciences of the United States of America, 97*(21), 11455–11459.

Geier, M. R., Trigg, M. E., & Merril, C. R. (1973). Fate of bacteriophage lambda in non-immune germ-free mice. *Nature, 246*(5430), 221–223.

Górski, A., Kniotek, M., Perkowska-Ptasińska, A., Mróz, A., Przerwa, A., Gorczyca, W., … Nowaczyk, M. (2005). Bacteriophages and transplantation tolerance. *Transplantation Proceedings, 38*(1), 331–333.

Grabowska, A. M., Jennings, R., Laing, P., Darsley, M., Jameson, C. L., & Swift, L., et al. (2000). Immunisation with phage displaying peptides representing single epitopes of the glycoprotein G can give rise to partial protective immunity to HSV-2. *Virology, 269*(0042-6822 (Print)), 47–53.

Gross, A. L., Gillespie, J. W., & Petrenko, V. A. (2016). Promiscuous tumor targeting phage proteins. *Protein Engineering Design and Selection*, 1–11.

Halpern, B. N. (1959). The Role and function of the reticulo-endothelial system in immunological processes. *King's College Publication*, 321–338.

Hamzeh-Mivehroud, M., Alizadeh, A. A., Morris, M. B., Church, W. B., & Dastmalchi, S. (2013). Phage display as a technology delivering on the promise of peptide drug discovery. *Drug Discovery Today, 18*(23–24), 1144–1157.

Hanahan, D., & Weinberg, R. A. (2000). The hallmarks of cancer. *Cell, 100*, 57–70.

Hanahan, D., & Weinberg, R. A. (2011). Hallmarks of cancer: The next generation. *Cell, 144*(5), 646–674.

Hashiguchi, S., Yamaguchi, Y., Takeuchi, O., Akira, S., & Sugimura, K. (2010). Immunological basis of M13 phage vaccine: Regulation under MyD88 and TLR9 signaling. *Biochemical and Biophysical Research Communications, 402*(1), 19–22.

Henry, K. A., Arbabi-Ghahroudi, M., & Scott, J. K. (2015). Beyond phage display: Non-traditional applications of the filamentous bacteriophage as a vaccine carrier, therapeutic biologic, and bioconjugation scaffold. *Frontiers in Microbiology, 6*, 755.

Inchley, C. J. (1969). The activity of mouse Kupffer cells following intravenous injection of T4 bacteriophage. *Clinical and Experimental Immunology, 5*, 173–187.

Ivanenkov, V. V., Felici, F., & Menon, A. G. (1999). Uptake and intracellular fate of phage display vectors in mammalian cells. *Biochimica et Biophysica Acta, 1448*, 450–462.

Jain, R. K. (1990). Physiological barriers to delivery of monoclonal antibodies and other macromolecules in tumors. *Cancer Research, 50*(3 SUPPL.), 814–819.

Jayanna, P. K., Bedi, D., Deinnocentes, P., Bird, R. C., & Petrenko, V. A. (2010a). Landscape phage ligands for PC3 prostate carcinoma cells. *Protein Engineering, Design and Selection, 23*(6), 423–430.

Jayanna, P. K., Bedi, D., Gillespie, J. W., DeInnocentes, P., Wang, T., Torchilin, V. P., … Petrenko, V. A. (2010b). Landscape phage fusion protein-mediated targeting of nanomedicines enhances their prostate tumor cell association and cytotoxic efficiency. *Nanomedicine: Nanotechnology, Biology and Medicine, 6*(4), 538–546.

Jayanna, P. K., Torchilin, V. P., & Petrenko, V. A. (2009). Liposomes targeted by fusion phage proteins. *Nanomedicine: Nanotechnology, Biology and Medicine, 5*(1), 83–89.

Jończyk, E., Kłak, M., Międzybrodzki, R., & Górski, A. (2011). The influence of external factors on bacteriophages–review. *Folia Microbiologica, 56*(3), 191–200.

Karimi, M., Mirshekari, H., Moosavi Basri, S. M., Bahrami, S., Moghoofei, M., & Hamblin, M. R. (2016). Bacteriophages and phage-inspired nanocarriers for targeted delivery of therapeutic cargos. *Advanced Drug Delivery Reviews*.

Kassner, P. D., Burg, M. A., Baird, A., & Larocca, D. (1999). Genetic selection of phage engineered for receptor-mediated gene transfer to mammalian cells. *Biochemical and Biophysical Research Communications, 264*(3), 921–928.

Kaur, T., Nafissi, N., Wasfi, O., Sheldon, K., Wettig, S., & Slavcev, R. (2012). Immunocompatibility of bacteriophages as nanomedicines. *Journal of Nanotechnology*.

Kim, Y., Caberoy, N. B., Alvarado, G., Davis, J. L., Feuer, W. J., & Li, W. (2011). Identification of Hnrph3 as an autoantigen for acute anterior uveitis. *Clinical Immunology (Orlando, Fla.),* *138*(1), 60–66.

Kolonin, M. G., Saha, P. K., Chan, L., Pasqualini, R., & Arap, W. (2004). Reversal of obesity by targeted ablation of adipose tissue. *Nature Medicine, 10*(6), 625–632.

Ladner, R. C., Sato, A. K., Gorzelany, J., & de Souza, M. (2004). Phage display-derived peptides as therapeutic alternatives to antibodies. *Drug Discovery Today, 9*(12), 525–529.

Larocca, D., Jensen-Pergakes, K., Burg, M. A., & Baird, A. (2001). Receptor-targeted gene delivery using multivalent phagemid particles. *Molecular Therapy, 3*(4), 476–484.

Lewis, V. O., Ozawa, M. G., Deavers, M. T., Wang, G., Shintani, T., Arap, W., et al. (2009). The interleukin-11 receptor alpha as a candidate ligand-directed target in osteosarcoma: Consistent data from cell lines, orthotopic models, and human tumor samples. *Cancer Research, 69*(5), 1995–1999.

Lopez, V., Ochs, H. D., Thuline, H. C., Davis, S. D., & Wedgwood, R. J. (1975). Defective antibody response to bacteriophage ØX 174 in down syndrome. *The Journal of Pediatrics, 86*(2), 207–211.

Luken, B. M., Kaijen, P. H. P., Turenhout, E. A. M., Kremer Hovinga, J. A., van Mourik, J. A., Fijnheer, R., et al. (2006). Multiple B-cell clones producing antibodies directed to the spacer and disintegrin/thrombospondin type-1 repeat 1 (TSP1) of ADAMTS13 in a patient with acquired thrombotic thrombocytopenic purpura. *Journal of Thrombosis and Haemostasis: JTH, 4*(11), 2355–2364.

March, J. B., Clark, J. R., & Jepson, C. D. (2004). Genetic immunisation against hepatitis B using whole bacteriophage λ particles. *Vaccine, 22,* 1666–1671.

Merril, C. R., Biswas, B., Carltont, R., Jensen, N. C., Creed, G. J., Zullo, S., et al. (1996). Long-circulating bacteriophage as antibacterial agents. *Proceedings of the National Academy of Sciences, 93,* 3188–3192.

Moghimi, S. M., Hunter, A. C., & Murray, J. C. (2005). Nanomedicine: Current status and future prospects. *FASEB Journal, 19*(3), 311–330.

Molek, P., Strukelj, B., & Bratkovic, T. (2011). Peptide phage display as a tool for drug discovery: Targeting membrane receptors. *Molecules (Basel, Switzerland), 16*(1), 857–87.

Molenaar, T. J. M., Michon, I., de Haas, S. A. M., van Berkel, T. J. C., Kuiper, J., & Biessen, E. A. L. (2002). Uptake and processing of modified bacteriophage M13 in mice: Implications for phage display. *Virology, 293*(1), 182–191.

Mount, J. D., Samoylova, T. I., Morrison, N. E., Cox, N. R., Baker, H. J., & Petrenko, V. A. (2004). Cell targeted phagemid rescued by preselected landscape phage. *Gene, 341,* 59–65.

Newton, J. R., Kelly, K. A., Mahmood, U., Weissleder, R., & Deutscher, S. L. (2006). In vivo selection of phage for the optical imaging of PC-3 human prostate carcinoma in mice. *Neoplasia, 8*(9), 772–780.

Niu, Z., Bruckman, M. A., Harp, B., Mello, C. M., & Wang, Q. (2008). Bacteriophage M13 as a scaffold for preparing conductive polymeric composite fibers. *Nano Research, 1*(3), 235–241.

Ochs, H. D., Davis, S. D., & Wedgwood, R. J. (1971). Immunologic responses to bacteriophage phi-X 174 in immunodeficiency diseases. *The Journal of Clinical Investigation, 50*(12), 2559–2568.

Olszowska-Zaremba, N., Borysowski, J., Dabrowska, K., & Gorski, A. (2012a). Phage translocation, safety and immunomodulation. In *Bacteriophages in health and disease* (pp. 168–184). Cambridge, MA: Advances in Molecular and Cellular Microbiology.

Olszowska-Zaremba, N., Borysowski, J., Dabrowska, K., Górski, A., Hyman, P., & Abedon, S. T. (2012b). Phage translocation, safety and immunomodulation. In P. Hyman & S. T. Abedon (Eds.), *Bacteriophages in health and disease* (pp. 168–184). Wallingford: CABI.

Pasqualini, R., Millikan, R. E., Christianson, D. R., Cardó-Vila, M., Driessen, W. H. P., Giordano, R. J., … Arap, W. (2015). Targeting the interleukin-11 receptor α in metastatic prostate cancer: A first-in-man study. *Cancer, 121*(14), 2411–2421.

Petrenko, V. A., & Jayanna, P. K. (2014). Phage protein-targeted cancer nanomedicines. *FEBS Letters, 588*(2), 341–349.

Poul, M. A., Becerril, B., Nielsen, U. B., Morisson, P., & Marks, J. D. (2000). Selection of tumor-specific internalizing human antibodies from phage libraries. *Journal of Molecular Biology, 301*(5), 1149–1161.

Poul, M. A., & Marks, J. D. (1999). Targeted gene delivery to mammalian cells by filamentous bacteriophage. *Journal of Molecular Biology, 288*(2), 203–211.

Rangel, R., Dobroff, A. S., Guzman-Rojas, L., Salmeron, C. C., Gelovani, J. G., Sidman, R. L., ... Arap, W. (2013). Targeting mammalian organelles with internalizing phage (iPhage) libraries. *Nature Protocols, 8*(10), 1916–1939.

Rangel, R., Guzman-Rojas, L., le Roux, L. G., Staquicini, F. I., Hosoya, H., Barbu, E. M., ... Arap, W. (2012). Combinatorial targeting and discovery of ligand-receptors in organelles of mammalian cells. *Nature Communications, 3*, 788.

Rao, A. J., Ramachandra, S. G., Ramesh, V., Couture, L., Abdennebi, L., Salesse, R., et al. (2004). Induction of infertility in adult male bonnet monkeys by immunization with phage-expressed peptides of the extracellular domain of FSH receptor. *Reproductive BioMedicine Online, 8*(4), 385–391.

Rishi, P., Singh, A. P., Arora, S., Garg, N., & Kaur, I. P. (2014). Revisiting eukaryotic anti-infective biotherapeutics. *Critical Reviews in Microbiology, 40*(4), 281–292.

Roehnisch, T., Then, C., Nagel, W., Blumenthal, C., Braciak, T., Donzeau, M., ... Oduncu, F. (2013). Chemically linked phage idiotype vaccination in the murine B cell lymphoma 1 model. *Journal of Translational Medicine, 11*(1), 267.

Roehnisch, T., Then, C., Nagel, W., Blumenthal, C., Braciak, T., Donzeau, M., ... Oduncu, F. S. (2014). Phage idiotype vaccination: first phase I/II clinical trial in patients with multiple myeloma. *Journal of Translational Medicine, 12*(1), 119. http://doi.org/10.1186/1479-5876-12-119

Rosenberg, S. A. (1999). A new era of cancer immunotherapy: Converting theory to performance. *CA: A Cancer Journal for Clinicians, 49*, 70–73.

Samoylov, A., Cochran, A., Schemera, B., Kutzler, M., Donovan, C., Petrenko, V., ... Samoylova, T. (2015). Humoral immune responses against gonadotropin releasing hormone elicited by immunization with phage-peptide constructs obtained via phage display. *Journal of Biotechnology, 216*, 20–28.

Samoylova, T. I., Petrenko, V. A., Morrison, N. E., Globa, L. P., Baker, H. J., & Cox, N. R. (2003). Phage probes for malignant glial cells. *Molecular Cancer Therapeutics, 2*(16), 1129–1137.

Sblattero, D., Berti, I., Trevisiol, C., Marzari, R., Bradbury, A., Not, T., ... Ventura, A. (1999). Human tissue transglutaminase ELISA: a powerful mass screening diagnostic assay for celiac disease. *Journal of Pediatric Gastroenterology and Nutrition, 28*(5), 568.

Smith, G. P., & Petrenko, V. A. (1997). Phage display. *Chemical Reviews, 2665*(96), 391–410.

Sokoloff, A. V., Bock, I., Zhang, G., Sebestyén, M. G., & Wolff, J. A. (2000). The interactions of peptides with the innate immune system studied with use of T7 phage peptide display. *Molecular Therapy, 2*(2), 131–139.

Srivastava, A. S., Kaido, T., & Carrier, E. (2004). Immunological factors that affect the in vivo fate of T7 phage in the mouse. *Journal of Virological Methods, 115*(1), 99–104.

Strebhardt, K., & Ullrich, A. (2008). Paul Ehrlich's magic bullet concept: 100 years of progress. *Nature Reviews Cancer, 8*(6), 473–480.

Sulakvelidze, A. (2005). Phage therapy: An attractive option for dealing with antibiotic-resistant bacterial infections. *Drug Discovery Today, 10*(12), 807–809.

Sulakvelidze, A., Alavidze, Z., & Glenn, J. M., Jr. (2001). Bacteriophage therapy. *Antimicrobial Agents and Chemotherapy, 45*(3), 649–659.

Trouche, S. G., Asuni, A., Rouland, S., Wisniewski, T., Frangione, B., Verdier, J. M., ... Mestre-Francés, N. (2009). Antibody response and plasma Aβ1-40 levels in young *Microcebus murinus* primates immunized with Aβ1-42 and its derivatives. *Vaccine, 27*(7), 957–964.

Ulivieri, C., Citro, A., Ivaldi, F., Mascolo, D., Ghittoni, R., Fanigliulo, D., ... Del Pozzo, G. (2008). Antigenic properties of HCMV peptides displayed by filamentous bacteriophages vs. synthetic peptides. *Immunology Letters, 119*(1–2), 62–70.

Urbanelli, L., Ronchini, C., Fontana, L., Menard, S., Orlandi, R., & Monaci, P. (2001). Targeted gene transduction of mammalian cells expressing the HER2/neu receptor by filamentous phage. *Journal of Molecular Biology, 313*, 965–976.

van Houten, N. E., Henry, K. A., Smith, G. P., & Scott, J. K. (2010). Engineering filamentous phage carriers to improve focusing of antibody responses against peptides. *Vaccine, 28*(10), 2174–2185.

van Houten, N. E., Zwick, M. B., Menendez, A., & Scott, J. K. (2006). Filamentous phage as an immunogenic carrier to elicit focused antibody responses against a synthetic peptide. *Vaccine, 24*(19), 4188–4200.

Wang, T., A. Petrenko, V., & Torchilin, V. P. (2011). Optimization of landscape phage fusion protein-modified polymeric Peg-Pe micelles for improved breast cancer cell targeting. *Journal of Nanomedicine & Nanotechnology, s4*(01), 008.

Watanabe, K., Goodrich, J., Li, C., Agha, M., & Greenberg, P. (1992). Restoration of viral immunity in immunodeficient humans by the adoptive transfer of T cell clones. *Science, 257*(5067), 238–241.

Willis, A. E., Perham, R. N., & Wraith, D. (1993). Immunological properties of foreign peptides in multiple display on a filamentous bacteriophage. *Gene, 128*(1), 79–83.

Wu, Y., Wan, Y., Bian, J., Zhao, J., Jia, Z., Zhou, L., ... Tan, Y. (2002). Phage display particles expressing tumor-specific antigens induce preventive and therapeutic anti-tumor immunity in murine p815 model. *International Journal of Cancer, 98*(5), 748–753.

Yang, W. J., Lai, J. F., Peng, K. C., Chiang, H. J., Weng, C. N., & Shiuan, D. (2005). Epitope mapping of *Mycoplasma hyopneumoniae* using phage displayed peptide libraries and the immune responses of the selected phagotopes. *Journal of Immunological Methods, 304*(1–2), 15–29.

Zuercher, A. W., Miescher, S. M., Vogel, M., Rudolf, M. P., Michael, B., & Stadler, B. M. (2004). Oral anti-IgE immunization with epitope-displaying phage. *Veterinary Immunology and Immunopathology, 99*, 11–24.

Zurita, A. J. (2004). Combinatorial screenings in patients: The interleukin-11 receptor as a candidate target in the progression of human prostate cancer. *Cancer Research, 64*(2), 435–439.

Printed in the United States
By Bookmasters